HYDRODYNAMICS OF OCEANS AND ATMOSPHERES

HYDRODYNAMICS OF OCEANS AND ATMOSPHERES

by

CARL ECKART
University of California
Scripps Institution of Oceanography

PERGAMON PRESS
OXFORD · LONDON · NEW YORK · PARIS
1960

PERGAMON PRESS LTD.
Headington Hill Hall, Oxford
4 & 5 Fitzroy Square, London W.1.

PERGAMON PRESS INC.
122 East 55th Street, New York 22, N.Y.
P.O. Box 47715, Los Angeles, California

PERGAMON PRESS S.A.R.L
24 Rue des Écoles, Paris V^e

PERGAMON PRESS G.m.b.H.
Kaiserstrasse 75, Frankfurt am Main

Library of Congress Card Number 60-9643

Printed in Northern Ireland at The Universities Press, Belfast

CONTENTS

CHAPTER IV. THE FIELD EQUATIONS

CHAPTER V. THE EARTH'S ATMOSPHERE, OCEANS AND LAKES

CHAPTER VI. GENERAL THEOREMS CONCERNING THE FIELD EQUATIONS

CHAPTER VII. FORMULATION OF THE MAJOR MATHEMATICAL PROBLEMS

CHAPTER VIII. THE ISOTHERMAL ATMOSPHERE: PLANE LEVEL SURFACES WITHOUT ROTATION

CHAPTER IX. THE ISOTHERMAL ATMOSPHERE: PLANE LEVEL SURFACES WITH ROTATION

CHAPTER X. OCEANS WITH CONSTANT COEFFICIENTS

PREFACE

THE greater part of this book could have been written fifty or sixty years ago. It is a systematic account of the simplest motions of a stratified compressible fluid. The known motions of the earth's oceans and atmosphere are not all of this simplicity. It is therefore inevitable that the book will seem anachronistic to many readers and will ignore many generally accepted facts. In some cases the author doubts the validity of the theories that have been used to establish these facts, but it has not been possible to give a detailed critical discussion of these points without abandoning the principal objective of the book.

For example, the following pages will contain no account of the effects of viscosity and turbulence, which modern meteorology and oceanography invoke so freely to explain many phenomena. The reason is that, in geophysics, there are two Reynolds numbers. The first is the ordinary one:

$$R = VL/\nu$$

V being a typical velocity, L a typical length, and ν the kinematic viscosity. The other is

$$R_g = NL^2/\nu$$

N being the Väisälä–Brunt frequency (see pages 59–61). In laboratory experiments, $R_g \ll R$, and the latter will dominate the phenomena. For large, or moderately large-scale phenomena in the geophysical fluids, $R_g \gg R$, and the former will dominate. Consequently it is most improbable that the results of laboratory investigations of turbulence will have much relevance to turbulence in the air or oceans. In particular, the theory of isotropic turbulence will have little application to those fluctuations of the wind that disturb the even flight of an aircraft.

However, the theory of geophysical turbulence belongs to the superstructure for which this book attempts to provide the foundation. It has therefore found no place in the following pages.

C. E.

La Jolla, March 15, 1960

CHAPTER I

THE BASIC EQUATIONS

1. Introduction

THE most striking phenomenon of the Earth's atmosphere is the diminution of its density with altitude. This stratification endows it with a stability that is completely lacking in a homogeneous fluid. Although it is not so well known, the earth's oceans are also stratified and have a stability of the same order as that of the atmosphere. On the other hand, classical hydrodynamics is concerned, almost exclusively, with the motion of a homogeneous, incompressible fluid. It has been conspicuously unsuccessful in explaining meteorological and oceanographic phenomena, presumably for this reason.

There has been no attempt to make a systematic theory of the motion of a stably stratified fluid. A review of the literature reveals more dubious assumptions than well-established results.(4) The objective of this book will be to derive fairly general equations, and to study their solutions systematically, using the standard techniques of wave theory. The objective will be to illustrate the laws of dynamics, rather than to imitate nature. The final step, of introducing empirical data into the calculations and comparing the result with observation, will usually not be attempted.

The derivation of the equations will necessarily be based on assumptions, and some of these will be drastic. The first concerns the nature of the fluid. Both air and sea-water are mixtures; this will be ignored, and it will be supposed that the thermodynamic relations appropriate to a chemically pure fluid are applicable.

Secondly, the non-linear equations of thermodynamics and hydrodynamics will be studied by the methods of perturbation theory. That is, only small departures from a state of static equilibrium will be considered, so that the equations to be solved are linear. This doubtless excludes some phenomena from the study. The extent to which this is an oversimplification is not known; the author is inclined to think that it is possible to treat a wider range of topics in this way than is usually supposed.

2. Thermodynamics

The state of a pure fluid can be specified by any two of the four variables v, p, θ, η, which are respectively its specific volume, pressure, temperature and entropy.[3, 6] Its internal energy, ϵ, (erg/g) is best expressed as a function of v and η. Then its equation of state can be obtained by eliminating η between the two fundamental equations

$$p = -\partial\epsilon/\partial v, \qquad \theta = \partial\epsilon/\partial\eta. \tag{1}$$

These can also be used to transform from one pair of independent variables to another.

On differentiating Eq. (2–1) the result may be written

$$\begin{aligned} \delta p &= -X\,\delta v + Y\,\delta\eta, \\ \delta\theta &= -Y\,\delta v + Z\,\delta\eta, \end{aligned} \tag{2}$$

where $+X$, $-Y$ and Z are the three second derivatives of ϵ. They can be expressed in terms of more commonplace quantities as follows:

$$\begin{aligned} X &= \rho^2 c^2, \\ Y &= \rho(\gamma - 1)/a, \\ Z &= \theta/C_v, \end{aligned} \tag{3}$$

where $\rho = 1/v =$ density,
$c =$ velocity of sound,
$a =$ coefficient of thermal expansion,
$C_v =$ specific heat, constant volume,
$\gamma =$ ratio of specific heats.

The derivation of these equations is as follows: The definition of the velocity of sound, as given in text books on acoustics,[9] may be reduced to

$$c^2 = \left(\frac{\delta p}{\delta \rho}\right)_{\delta\eta = 0}.$$

Since $\delta v = -\delta\rho/\rho^2$, the first of Eqs. (2–1) immediately yields

$X = \rho^2 c^2$. It will be noted that ρc is the acoustic impedance of the fluid.[9] In the same way, it is a matter of definition that

$$C_v = \left(\frac{\delta \epsilon}{\delta \theta}\right)_{\delta v = 0} = \theta \left(\frac{\delta \eta}{\delta \theta}\right)_{\delta v = 0},$$

whereupon it follows at once that $Z = \theta / C_v$. To establish the formula for Y, it is first necessary to calculate the specific heat at constant pressure, defined by

$$C_p = \left(\frac{\delta \epsilon + p\, \delta v}{\delta \theta}\right)_{\delta p = 0} = \theta \left(\frac{\delta \eta}{\delta \theta}\right)_{\delta p = 0}.$$

When $\qquad \delta p = 0, \qquad \delta v = (Y/X)\, \delta \eta,$

and so

$$\delta \theta = \left(1 - \frac{Y^2}{XZ}\right) Z\, \delta \eta.$$

From this, it follows that

$$C_p = \left(1 - \frac{Y^2}{XZ}\right)^{-1} C_v,$$

or

$$C_v / C_p = 1/\gamma = 1 - \frac{Y^2}{XZ}.$$

Consequently

$$Y^2 = \frac{\gamma - 1}{\gamma} XZ = \frac{(\gamma - 1)\rho^2 c^2 \theta}{\gamma C_v}. \tag{4}$$

This equation would enable one to determine Y (except for sign) without reference to the coefficient of thermal expansion. This last is defined as

$$a = \frac{1}{v}\left(\frac{\delta v}{\delta \theta}\right)_{\delta p = 0}.$$

Using the foregoing expression for γ, this leads to $Y = \rho(\gamma - 1)/a$, as stated above. Equation (2–4) may then be transformed to

$$\gamma(\gamma - 1) C_v = a^2 c^2 \theta, \tag{5}$$

which is a thermodynamic identity of some importance.

3. The thermodynamic functions of an ideal gas

As a more concrete example of the general thermodynamic relations, and one that will be useful later, the case of an ideal gas may be considered. An ideal gas is one whose molecules are all identical and interact only by collisions, moving in straight lines between collisions. Its specific heats are constants. The general principles of statistical mechanics(7) can then be used to show that its internal energy

$$\epsilon = Av^{-\gamma+1} \exp [(\gamma - 1)\eta/R], \tag{1}$$

A, γ and R being constants.

In particular, γ is the ratio of specific heats, as defined above. For air of molecular weight 29, $R = 2{\cdot}87 \times 10^6$ erg/g deg and $\gamma = 1{\cdot}40$. The value of A will not be needed.

Then it follows from Eq. (2–1) that

$$p = -\frac{\partial \epsilon}{\partial v} = (\gamma - 1)Av^{-\gamma} \exp [(\gamma - 1)\eta/R], \tag{2}$$

$$\theta = \frac{\partial \epsilon}{\partial \eta} = [(\gamma - 1)A/R]v^{-\gamma+1} \exp [(\gamma - 1)\eta/R].$$

These equations may be transformed into

$$\begin{aligned} pv &= R\theta, \\ pv^{\gamma} &= (\gamma - 1)A \exp [(\gamma - 1)\eta/R], \end{aligned} \tag{3}$$

which are the *equation of state* and the *equation of the isentropes*, respectively.

The second derivative of ϵ yields

$$\begin{aligned} X &= \frac{\partial^2 \epsilon}{\partial v^2} = \gamma(\gamma - 1)Av^{-\gamma-1} \exp [(\gamma - 1)\eta/R] \\ &= \gamma\rho^2 R\theta. \end{aligned} \tag{4}$$

Combined with the previous result $X = \rho^2 c^2$, it follows that the velocity of sound, c, is given by Laplace's formula:

$$c^2 = \gamma R\theta. \tag{5}$$

Similarly

$$\begin{aligned} +Y &= -\frac{\partial^2 \epsilon}{\partial v\, \partial \eta} = [(\gamma - 1)^2/R]Av^{-\gamma} \exp [(\gamma - 1)\eta/R] \\ &= (\gamma - 1)\rho\theta. \end{aligned} \tag{6}$$

Combined with $Y = \rho(\gamma - 1)/a$, this shows that the coefficient of thermal expansion is

$$a = 1/\theta. \tag{7}$$

Finally

$$Z = \frac{\partial^2 \epsilon}{\partial \eta^2} = [(\gamma - 1)/R]^2 A v^{-\gamma+1} \exp[(\gamma - 1)\eta/R]$$

$$= (\gamma - 1)\theta/R. \tag{8}$$

From $Z = \theta/C_v$, one finds that the specific heat is

$$C_v = R/(\gamma - 1). \tag{9}$$

Substitution of Eqs. (3–5), (3–7) and (3–9) into Eq. (2–5) verifies that the latter is satisfied.

4. The thermodynamic coefficients for water

It would be pleasant to assume that the basic coefficients of fresh water and sea-water are constants. This would amount to supposing

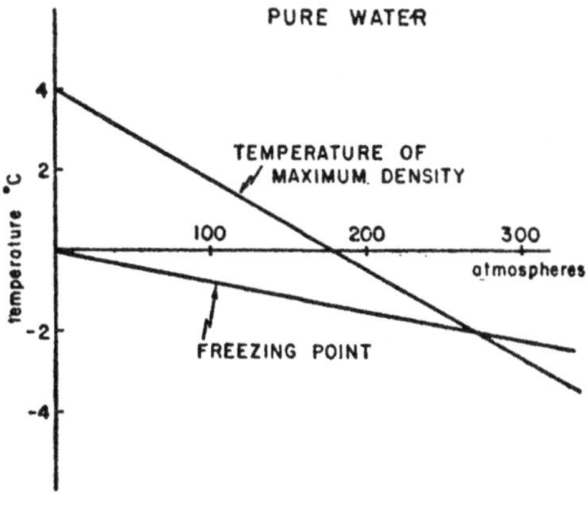

FIG. 1. The variation with pressure of the freezing point and the temperature of maximum density of pure water.

that, in the p–v diagram, both the *isotherms* and *adiabatics* can be approximated by uniformly-spaced parallel straight lines. Unfortunately, the phenomenon of *maximum density* cannot be ignored in the study of fresh-water lakes; and even in salt-water, the isotherms are markedly curved, though maximum density is not reached except in the super-cooled state (see Figs. 1 and 2).

Specifically, the coefficient of thermal expansion of these fluids is not a constant, but vanishes at a certain temperature, which is itself a function of pressure. In turn, the thermodynamic identity

$$\gamma(\gamma - 1)C_v = a^2c^2\theta$$

shows that $\gamma - 1$ has a double root at this temperature and hence that Y has a simple root. Thus, the simplest permissible assumption requires Y to be a linear function of pressure and temperature.

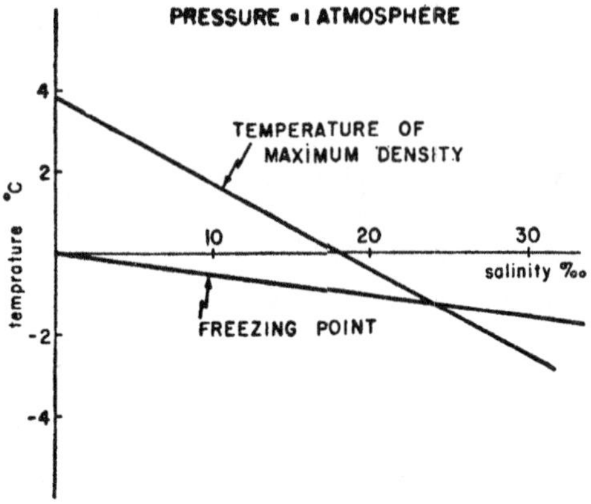

Fig. 2. The variation with salinity of the freezing point and the temperature of maximum density of sea water at 1 atmosphere pressure.

For pure water, such as is found in lakes, the temperature of maximum density may be supposed to be $4 - 2 \times 10^{-8}p$ °C, although no great precision can be claimed for this formula. It is convenient to introduce the quantity

$$T_M = T - 4{\cdot}0 + 2 \times 10^{-8}p, \tag{1}$$

T being the temperature (°C). Then, in the range from 0 to 7°C, and 0 to 20 atm, the coefficient of thermal expansion is approximately

$$a = 15 \times 10^{-6}T_M, \tag{2}$$

and the ratio of the specific heats is

$$\gamma = 1 + 3{\cdot}0 \times 10^{-5}T_M{}^2. \tag{3}$$

The three basic coefficients X, Y, Z are then approximately

$$X = 2{\cdot}0 \times 10^{10},$$
$$Y = 2{\cdot}0 T_M, \qquad (4)$$
$$Z = 6{\cdot}6 \times 10^{-6},$$

in c.g.s. units. To this approximation

$$(\rho c)^{1/2} = X^{1/4} = 375. \qquad (5)$$

In this book, the velocity of sound will usually be considered as constant; however, a somewhat more precise formula is

$$c = 1{\cdot}42 \times 10^5[1 + 35 \times 10^{-4}T_M + 2 \times 10^{-9}p] \text{ cm/sec.} \qquad (6)$$

For sea-water, of salinity 35 parts per thousand, and in the ranges 0–15°C, 0–1000 atm, the corresponding formulae are

$$T_M = T + 7{\cdot}0 + 2 \times 10^{-9}p, \qquad (7)$$
$$a = 9{\cdot}3 \times 10^{-6}T_M, \qquad (8)$$
$$\gamma = 1 + 1{\cdot}2 \times 10^{-5}T_M{}^2, \qquad (9)$$
$$X = 2{\cdot}1 \times 10^{10}, \qquad (10)$$
$$Y = 1{\cdot}3T_M, \qquad (11)$$
$$Z = 6{\cdot}7 \times 10^{-6}. \qquad (12)$$

It will be noted that Eq. (4–7) does not yield the same temperature of maximum density as does Fig. 1. The discrepancy arises because the constants have been adapted to obtain the best fit over the whole pressure range from 0 to 1000 atm, and this results in a considerable error at the extremes of this range.

These expressions for the thermodynamic properties of pure water and sea-water are very approximate. Ultimately, they should be replaced by more exact expressions.[5] They do, however, represent the properties of these fluids more realistically than would the assumption that their coefficients of compressibility, thermal expansion, etc. are independent of temperature and pressure.

5. Hydrodynamics

The derivation of the hydrodynamic equations is discussed in Refs. 1, 3, 8 and will not be given here. The equations are

$$\frac{D\mathbf{u}}{Dt} + v\,\nabla p + g\,\nabla\chi + \mathbf{\Omega} \times \mathbf{u} = \mathbf{f}, \tag{1}$$

$$\frac{Dv}{Dt} - v\,\nabla\cdot\mathbf{u} = 0, \tag{2}$$

$$\frac{D\eta}{Dt} = \frac{q}{\theta}, \tag{3}$$

where p, v, η, θ retain the meanings given them above, and

$\mathbf{u}$ = velocity (cm/sec);
$g\chi$ = gravitational potential;
g = 980 (cm/sec^2), by international convention;
χ = "gravitational altitude," (cm);
$\mathbf{\Omega}$ = Coriolis vector (rad/sec);
$\mathbf{f}$ = resultant of all other forces, (dyne/g);
q = net accession of heat, (erg/g sec);
$D/Dt = \partial/\partial t + \mathbf{u}\cdot\nabla$ = the Eulerian derivative.

Gravitational potential. The "gravitational altitude," χ, may, to an accuracy sufficient for the present purposes, be considered equal to the vertical co-ordinate, or to differ from it only by an additive constant. If the level surfaces are approximated by planes, $\chi = z$, where z is the vertical cartesian co-ordinate. If the level surfaces are approximated by spheres, $\chi = r - r_s$, where r is the distance of the point from the centre of the Earth, and r_s the radius of the solid earth (or mean sea-level).

Rotation. The Coriolis vector has the direction of the Earth's axis of rotation and points toward Polaris. Its magnitude is given by $2\pi/\Omega$ = 12 sidereal hours, or $\Omega = 1{\cdot}45 \times 10^{-4}$ rad/sec. (cf. Section 30).

Forces. The force $\mathbf{f}$ includes several components. The most important of these is the tide-producing force (Ref. 8, p. 358). This is, to a high degree of approximation, independent of the velocity and thermodynamic variables; it can therefore be considered as a known function of time and space. Another component results from the viscous stresses, but this is usually quite negligible in large scale geophysical problems.

Accession of heat. The net accession of heat is due to three causes: heat conduction, radiation and the viscous degradation of energy. The third component is quite negligible. The first component is given by

$$q = v \nabla \cdot (\sigma \nabla \theta),$$

σ being the thermal conductivity of the fluid. While this form for q has frequently been studied, it is presumably very much smaller than the contributions from solar radiation, and the back-radiation of infra-red to the sky. The theory of the radiative transfer of heat is not simple and will not be discussed here.[2] Instead q, like $\mathbf{f}$, will be treated as a known function of time and space.

Advection. Equations (5–1) to (5–3) contain the Eulerian derivative, $\mathrm{D}/\mathrm{D}t$. The significance of this derivative, as opposed to the partial derivative $\partial/\partial t$, deserves comment. Any quantity, such as velocity, density, temperature, etc. may be ascribed either to a bit of the fluid or to the point x, y, z that is momentarily occupied by that bit of fluid. The former is the more fundamental description. The rate at which the temperature of a bit of fluid is changing is, by definition, $\mathrm{D}\theta/\mathrm{D}t$. The rate at which the temperature of the fluid at x, y, z is changing is, again by definition, $\partial\theta/\partial t$. The difference between the two arises because different bits of fluid occupy the point x, y, z in succession, and can be shown to be

$$\frac{\partial \theta}{\partial t} - \frac{\mathrm{D}\theta}{\mathrm{D}t} = -\mathbf{u} \cdot \nabla\theta, \quad \text{etc.}$$

This succession of different bits of fluid at a single point is called advection and $-\mathbf{u} \,.\, \nabla\theta$, the rate of increase of θ at x, y, z due to advection. If the temperature of each bit of the fluid remains constant (though different bits have different temperatures), advection is the only cause of the change in temperature at any point. In this discussion, temperature has been used only as an example, and the same terminology can be applied to any other numerical property of the fluid.

Elimination of entropy. Equation (5–3) can be given an alternative form that apparently avoids the entropy concept. Because of Eq. (2–2),

$$\frac{D\eta}{Dt} = \frac{1}{Y}\left(\frac{Dp}{Dt} + X\frac{Dv}{Dt}\right)$$

$$= \frac{1}{Y}\left(\frac{Dp}{Dt} - c^2\frac{D\rho}{Dt}\right),$$

so that Eq. (5–3) is equivalent to

$$\frac{Dp}{Dt} - c^2\frac{D\rho}{Dt} = \frac{\rho a c^2}{C_p} q. \tag{4}$$

This form of the equation is frequently found in the literature.

Conservation of energy. Equations (5–1), (5–2) and (5–3) have an important consequence. If Eq. (5–1) is multiplied term by term by $\rho\mathbf{u}\cdot$, the result is

$$\tfrac{1}{2}\rho\frac{D\mathbf{u}^2}{Dt} + \mathbf{u}\cdot\nabla p = \rho\mathbf{u}\cdot\mathbf{f} + \rho g\mathbf{u}\cdot\nabla\chi.$$

Using Eq. (2–1) and Eqs. (5–2), (5–3), one obtains the first law of thermodynamics:

$$\rho\frac{D\epsilon}{Dt} = -p\,\nabla\cdot\mathbf{u} + \rho q. \tag{5}$$

Adding these two equations and noting that

$$\frac{D\chi}{Dt} = \mathbf{u}\cdot\nabla\chi,$$

one obtains

$$\rho\frac{D}{Dt}\{\tfrac{1}{2}\mathbf{u}^2 + \epsilon + g\chi\} + \nabla\cdot(p\mathbf{u}) = \rho\mathbf{u}\cdot\mathbf{f} + \rho q. \tag{6}$$

This is the equation that expresses the conservation of energy, $(\frac{1}{2})\mathbf{u}^2 + \epsilon + g\chi$ being the energy density (erg/g), and $p\mathbf{u}$ the energy flux (erg/cm² sec).

References for Chapter I

1. V. Bjerknes, J. Bjerknes, H. Solberg and T. Bergeron, *Physikalische Hydrodynamik*, Springer, Berlin, 1933.

2. S. Chandrasekhar, *Radiative Transfer*, Oxford Univ. Press, 470 pp., 1950.
3. C. Eckart, Thermodynamics of irreversible processes, *Phys. Rev.* **58,** 267–275 (1940).
4. C. Eckart and H. G. Ferris, Equations of motion of the ocean and atmosphere, *Rev. Mod. Phys.* **23,** 48–52 (1956).
5. C. Eckart, The equation of state of water and sea water at low temperatures and pressures, *Am. J. Sci.* **256,** 225–240 (1958).
6. J. W. Gibbs, *Scientific Papers: Vol. I, Thermodynamics*, Longmans, Green & Co., New York, 434 pp., 1906.
7. J. W. Gibbs, *Elementary Principles in Statistical Mechanics*, Yale Univ. Press, New Haven, 207 pp., 1914.
8. H. Lamb, *Hydrodynamics*, Cambridge Univ. Press, 6th ed., 728 pp., 1932.
9. P. W. Morse, *Vibration and Sound*, McGraw-Hill, New York, 385 pp., 1948.

CHAPTER II

THE PERTURBATION EQUATIONS

6. Introduction

The perturbation method of solving complicated differential equations has been developed largely for the ordinary differential equations of astronomy. In this field, it has been used with marked success. The differential equations of hydrodynamics are partial, but the general ideas of perturbation theory remain applicable. They have been explicitly formulated in Chapter VII of Ref. 1. They have also been used, with success, in dealing with the equations of quantum theory.

It is often said that they do not lead to useful results in hydrodynamics. Two reasons for this may be considered: either the basic equations of hydrodynamics may be wrong, or the perturbation method may have been used with inadequate care. A survey of the literature[2] tends to support the second surmise. In any event, it is premature to accept the first alternative without a much more thorough investigation than has yet been made.

The following pages contain the beginnings of such an investigation. In general, only the first two of the successive approximations of perturbation theory are considered. In the terminology that has been adopted in other fields, these are the zeroth and the first-order terms of an infinite series. There is little doubt that second-order terms are also needed to describe the phenomena adequately. However, it is unlikely that much progress can be made with the second order until the zero and first-order calculations have been fully understood.

7. The zero-order solution

Consider a static condition of the fluid, in which $\mathbf{u} = 0$, and all quantities are independent of the time, t. Moreover, let $\mathbf{f} = 0$, and

$q = 0$. Equations (5–1, 2, 3) then reduce to the single vector equation

$$v \nabla p = -g \nabla \chi, \tag{1}$$

which has the solution

$$p = p_0(\chi),$$
$$\frac{1}{v} = \rho = -p_0'(\chi)/g, \tag{2}$$

p_0 being a positive, monotonically decreasing function of χ, and the accent indicating differentiation with respect to χ. Equation (7–2) is

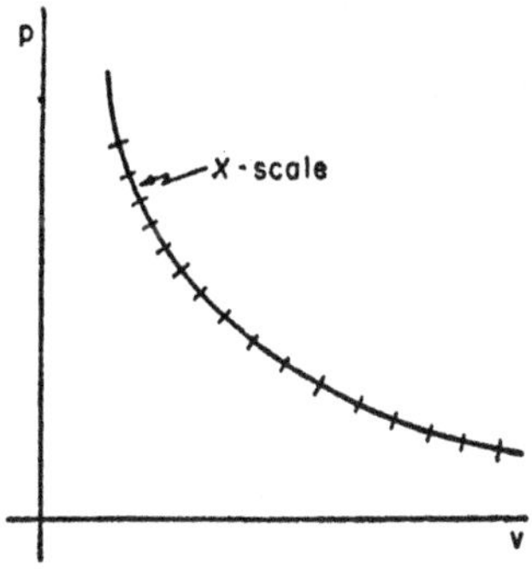

FIG. 3. Illustrating the concept of a barotropic state.

called the hydrostatic equation. The two quantities p and v suffice to determine the quantities θ and η, by Eq. (2–1). These are therefore also functions of χ alone, as are X, Y, Z. In general, these static values of any quantity will be called its zero-order values, and indicated by the subscript zero.

Barotropic states. Whenever pressure and density are both functions only of gravitational altitude, one speaks of a barotropic state. Barotropic states are quite special, in that any one of the three quantities p, v or χ determines the others. A convenient graphical representation is obtained by plotting p and v as rectangular co-ordinates; then the barotropic state is represented by a curve on this diagram, on which the scale of gravitational altitude may be marked (Fig. 3).

This is in marked contrast to the general case, in which p, v and χ are different functions of the co-ordinates of space and time, and may vary independently in the p–v diagram. A non-barotropic state would thus be represented by an area in the p–v diagram.

For most purposes, the p–η diagram is a more convenient way of showing the relations existing in a variable state of motion (see Section 47). This corresponds to the use of the pressure–potential temperature diagram in meteorology.

The gradients. Equation (2–2) leads to the identities

$$p_0' = -X_0 v_0' + Y_0 \eta_0',$$
$$\theta_0' = -Y_0 v_0' + Z_0 \eta_0', \tag{3}$$

which will be frequently needed. Using Eq. (7–2), the first of these becomes

$$\eta_0' = (X_0 v_0' - \rho_0 g)/Y_0$$

or, after using Eq. (2–3) for X_0 and Y_0,

$$\eta_0' = -\frac{a_0 c_0^2}{\gamma_0 - 1}\left(\frac{\rho_0'}{\rho_0} + \frac{g}{c_0^2}\right). \tag{4}$$

Because of Eq. (2–5)

$$\frac{a_0 c_0^2}{\gamma_0 - 1} = \frac{\gamma_0 C_{v0}}{a_0 \theta_0} = \frac{C_{p0}}{a_0 \theta_0}, \tag{5}$$

a formula that is often more convenient.

In much the same way, it can be shown that the temperature and density gradients are related by the equation

$$\theta_0' = -\frac{1}{a_0}\left(\frac{\rho_0'}{\rho_0} + \frac{\gamma_0 g}{c_0^2}\right), \tag{6}$$

and hence that

$$\theta_0' - \frac{\theta_0 \eta_0'}{C_{p0}} = -\frac{(\gamma_0 - 1)g}{a_0 c_0^2}. \tag{7}$$

If the entropy gradient vanishes, one speaks of adiabatic stratification; in this case,

$$\theta_0' = -\frac{(\gamma_0 - 1)g}{a_0 c_0^2} = -\frac{g a_0 \theta_0}{C_{p0}} = -\theta_A'. \tag{8}$$

The quantity θ_A' is called the adiabatic temperature gradient. In general Eq. (7–7) may therefore be written

$$\eta_0' = \frac{C_{p0}}{\theta_0}(\theta_0' + \theta_A'). \tag{9}$$

8. Zero-order states of an ideal gas

It is frequently desirable to use the example of an atmosphere of ideal gas since air at lower altitudes behaves much like one. The possible zero-order states are infinite in number, but a few special cases deserve mention.

The isothermal atmosphere. This was used as an example by Laplace, in his *Mécanique Céleste*(4) Vol. II. The barometric equation then integrates to

$$p_0 = p_s \exp(-g\chi/R\theta_0), \tag{1}$$

p_s being the value of p_0 at $\chi = 0$, i.e. at the surface of the solid earth. Boyle's law yields

$$\rho_0 = \rho_s \exp(-g\chi/R\theta_0), \tag{2}$$

and the equation of entropy becomes

$$\eta_0 = g\chi/\theta_0 + \eta_s. \tag{3}$$

Hence the entropy gradient is $\eta_0' = g/\theta R_0 =$ constant, greater than zero. The quantity,

$$H = R\theta_0/g, \tag{4}$$

is a constant in this case and is called the scale height of the isothermal atmosphere. Its value is 8·0 km at 273°K, 8·6 km at 293°.

Each of the thermodynamic coefficients has a characteristic dependence on altitude:

$$\begin{aligned} X_0 &= X_s \exp(-2\chi/H), \\ Y_0 &= Y_s \exp(-\chi/H), \\ Z_0 &= Z_s. \end{aligned} \tag{5}$$

The isentropic atmosphere. The example of an atmosphere in which the entropy is independent of height has been used by many writers. It is often said that convection tends to produce such a zero-order state, but this is very doubtful.

The adiabatic relation

$$p_0 v_0^{\gamma} = \text{positive constant}, \tag{6}$$

when introduced into the barometric equation, integrates to

$$p_0^{(1-\gamma)/\gamma} = -b\chi + \text{const}, \qquad b > 0. \tag{7}$$

Combined with Boyle's Law and the barometric equation, this yields

$$R\theta_0 = R\theta_s - b\chi, \tag{8}$$

θ_s being the surface temperature. Consequently the absolute temperature, θ_0, becomes zero at the altitude $R\theta_s/b$, and negative at higher altitudes. It is therefore seen that this example will lead to unnecessary paradoxes.

Atmosphere with constant temperature gradient. If the temperature is to vary linearly with altitude, its gradient can only be positive, unless these same paradoxes are admitted. Let

$$\theta_0 = g\chi/R\nu + \theta_s \tag{9}$$

where ν is a positive constant. For a temperature gradient of 1° per km, $\nu = 34$. The scale height is then a linear function of χ:

$$H = R\theta_0/g = (\chi + \chi_s)/\nu \tag{10}$$

χ_s being a positive constant. For this reason, some writers use H instead of θ_0 or χ in their equations; this is a simplification only in this one example, as will be seen below.

Using Eq. (8–9), the barometric equation integrates to

$$\begin{aligned} p_0 &= p_s[\chi_s/(\chi_s + \chi)]^{\nu}, \\ \rho_0 &= \rho_s[\chi_s/(\chi_s + \chi)]^{\nu+1}, \end{aligned} \tag{11}$$

and the entropy equation becomes

$$\eta_0 = R[\nu + \gamma/(\gamma - 1)] \log (\chi_s + \chi) + \text{const.} \tag{12}$$

The entropy gradient is therefore inversely proportional to $\chi + \chi_s$ or to θ_0:

$$\eta_0' = \left[1 + \frac{\gamma}{\nu(\gamma - 1)}\right] \frac{g}{\theta_0}.$$

The three thermodynamic coefficients are

$$\begin{aligned} X_0 &= X_s[\chi_s/(\chi_s + \chi)]^{2\nu+1}, \\ Y_0 &= Y_s[\chi_s/(\chi_s + \chi)]^{\nu}, \\ Z_0 &= Z_s[(\chi_s + \chi)/\chi_s]. \end{aligned} \tag{13}$$

General case. The barometric equation shows that in the most general case, p_0 must be a positive, monotonically decreasing, function of χ, and leads to the equation

$$H = R\theta_0/g = -p_0/p_0'. \tag{14}$$

In the actual atmosphere (see Section 29) H is not a monotonic function of altitude and therefore is useless as a co-ordinate; it is simply a measure of temperature. There is no good reason to use two units of temperature, and therefore the scale height will not be used in the following equations.

The remaining thermodynamic functions are easily expressed in terms of p_0 and p_0', but will not be needed in explicit form.

9. The first-order equations

Consider next a state of the fluid in which $\mathbf{u} = \mathbf{u}_1$, $v = v_0 + v_1$, $p = p_0 + p_1$, etc. Substitution of these expressions in Eq. (2–1) and expansion by Taylor's theorem results in

$$\begin{aligned} p_1 &= -X_0 v_1 + Y_0 \eta_1 + \dots, \\ \theta_1 &= -Y_0 v_1 + Z_0 \eta_1 + \dots, \end{aligned} \tag{1}$$

where the dots indicate terms of higher degree in v_1, η_1. It will be supposed that $\mathbf{u}_1$, v_1, p_1 are, in some sense, small so that it is justified to neglect these terms of higher degree.

Equations (5–1, 2, 3) must also be approximated, with the systematic neglect of the terms of second and higher degrees in smaller quantities. Since $\mathbf{u}_1$ is small, the advective increase of any quantity $w = w_0 + w_1 + \dots$ is simply

$$-\mathbf{u}_1 \cdot \nabla w_0 = -w_0'\mathbf{u}_1 \cdot \nabla\chi = -w_0'\mathbf{u}_1 \cdot \boldsymbol{\zeta}$$

where $\boldsymbol{\zeta} = \nabla\chi$. The same approximation yields

$$\frac{\partial w}{\partial t} = \frac{\partial w_1}{\partial t},$$

since w_0 is independent of t. Consequently

$$\frac{\mathrm{D}w}{\mathrm{D}t} = \frac{\partial w_1}{\partial t} + w_0'\mathbf{u}_1 \, . \, \boldsymbol{\zeta} + \dots .$$

It is important to note that the advective rates of change are of first order, proportional to the zero-order gradients. These are the terms that cause the theory of a stratified fluid to be different from that of a homogeneous fluid, and they must therefore all be retained in the first-order equations.

Using this general result, the Eqs. (5–1, 2, 3) yield

$$\frac{\partial \mathbf{u}_1}{\partial t} + v_0 \nabla p_1 + v_1 p_0' \boldsymbol{\zeta} + \boldsymbol{\Omega} \times \mathbf{u}_1 = \mathbf{f}_1, \tag{2}$$

$$\frac{\partial v_1}{\partial t} + v_0' \mathbf{u}_1 \cdot \boldsymbol{\zeta} - v_0 \nabla \cdot \mathbf{u}_1 = 0, \tag{3}$$

$$\frac{\partial \eta_1}{\partial t} + \eta_0' \mathbf{u}_1 \cdot \boldsymbol{\zeta} = q_1/\theta_0. \tag{4}$$

Using Eq. (7–3) and Eq. (9–1), it then follows that

$$\frac{\partial p_1}{\partial t} + p_0' \mathbf{u}_1 \cdot \boldsymbol{\zeta} + X_0 v_0 \nabla \cdot \mathbf{u}_1 = Y_0 q_1/\theta_0, \tag{5}$$

$$\frac{\partial \theta_1}{\partial t} + \theta_0' \mathbf{u}_1 \cdot \boldsymbol{\zeta} + Y_0 v_0 \nabla \cdot \mathbf{u}_1 = Z_0 q_1/\theta_0. \tag{6}$$

These are the first-order equations of motion; only two of the last four equations are independent. It is usually convenient to work with Eqs. (9–2, 4, 5) although Eqs. (9–2, 3, 4) are a useful set in special cases (Chapter III).

Interpretation. The interpretation of Eqs. (9–2, 3, 4) is clear: these express the Newtonian law of motion, the conservation of mass, and the second law of thermodynamics, respectively. In each case, of course, second-order terms have been neglected. Equations (9–5, 6) are derivative and have no similar basic interpretation.

Using Eqs. (2–3, 5), Eq. (9–5) may be written

$$\frac{1}{\rho_0 c_0^2}\left(\frac{\partial p_1}{\partial t} + p_0' \mathbf{u}_1 \cdot \boldsymbol{\zeta}\right) + \nabla \cdot \mathbf{u}_1 = a_0 q_1/C_{p0}. \tag{7}$$

The quantity q_1/C_{p0} would be the rate of increase in the temperature of a bit of the fluid, if the heating took place at constant pressure. Consequently, $a_0 q_1/C_{p0}$ would be the increase in specific volume under these conditions. The factor $K_0 = 1/\rho_0 c_0^2$ is recognized as the adiabatic compressibility, and the parenthesis is $\mathrm{D}p/\mathrm{D}t$. Hence this equation states that the divergence of the velocity is equal to the thermal expansion diminished by the compression.

Similarly, Eq. (9–6) may be written

$$\frac{\partial \theta_1}{\partial t} + \theta_0{}'\mathbf{u}_1 \cdot \boldsymbol{\zeta} + \frac{\gamma_0 - 1}{a_0} \nabla \cdot \mathbf{u}_1 = q_1/C_{v0}. \tag{8}$$

The right side of this equation would be the rate of increase of temperature if the heating took place at constant volume; $(\gamma_0 - 1) \nabla \cdot \mathbf{u}_1/a_0$ is the rate of cooling because of expansion. This furnishes the required interpretation of this equation.

Vertical displacement. Equation (9–4) may be given another form by introducing the vertical displacement, h, of the fluid from its zero-order position. This quantity satisfies the equation

$$\frac{\mathrm{D}h}{\mathrm{D}t} = \mathbf{u} \cdot \boldsymbol{\zeta}. \tag{9}$$

By definition, $h_0 = 0$, so that $h = h_1 + \dots$, and the first approximation to Eq. (9–9) is

$$\frac{\partial h_1}{\partial t} = \mathbf{u}_1 \cdot \boldsymbol{\zeta}. \tag{10}$$

Hence Eq. (9–4) becomes

$$\frac{\partial}{\partial t} (\eta_1 + h_1 \eta_0{}') = q_1/\theta_0, \tag{11}$$

and indicates a close connection between heating and vertical displacement. The resulting phenomena are called convection. If the net accession of heat is zero, Eq. (9–11) integrates to

$$\eta_1 = -h_1 \eta_0{}', \tag{12}$$

an equation that will be useful below.

10. The additive barotropic terms

The first-order equations are quite elaborate, and their solution is a major problem. However, they have one very simple solution, trivial in a physical sense, but of considerable technical importance. This is the static solution, for which $\mathbf{u}_1$, $\mathbf{f}_1$, q_1, and all time-derivatives vanish identically. As in the case of the rigorous equations, all of the first-order equations then become identities, except the first. This reduces to

$$v_0 \nabla p_1 - v_1 p_0' \boldsymbol{\zeta} = 0. \tag{1}$$

Since

$$v_1/v_0 = -\rho_1/\rho_0, \tag{2}$$

Equation (10–1) is equivalent to

$$\nabla p_1 + (\rho_1 p_0'/\rho_0)\boldsymbol{\zeta} = 0; \tag{3}$$

moreover,

$$p_0' = -\rho_0 g, \tag{4}$$

so that

$$\nabla p_1 = \rho_1 g \nabla \chi, \tag{5}$$

which is formally identical with the zero-order equation, and is satisfied when p_1 is an arbitrary function of gravitational altitude, and

$$\rho_1 = -p_1'/g. \tag{6}$$

This is again the hydrostatic equation. Using Eq. (9–1) it is seen that not only ρ_1 and v_1 but also θ_1 and η_1 are determined once $p_1(\chi)$ is specified.

Moreover, the distribution of the thermodynamic variables is again barotropic: this static solution merely represents a modification of the zero-order state.

Now, suppose that any solution of the first-order equations has been found. Since these equations are linear, another solution will be obtained if an arbitrary barotropic solution, of the kind just discussed, is added to the given solution. The barotropic solution functions in much the same way as an additive constant of integration, and often has the same importance as such a constant. In other cases, again like a constant of integration, the additive barotropic terms may be arbitrarily set equal to zero without essential loss of generality.

11. Boundary conditions

Not every solution of the first-order equations will be acceptable, since the fluid will in general have boundaries. These may be solid surfaces, such as the air–ground interface, or free surfaces, such as the air–sea interface. Solid boundaries may be idealized as rigid; if $\mathbf{n}$ is the unit normal to such a surface, then

$$\mathbf{n} \cdot \mathbf{u}_1 = 0 \tag{1}$$

at every point of that surface.

The air–sea interface may be idealized, for present purposes, by supposing that the atmospheric pressure is constant at all of its points. This will, however, be applied only to the water; the air over the sea will not be treated in the following pages. To zero approximation, therefore, the free surface of the ocean will be level—say the surface $\chi = 0$. To the first approximation, the equation of the free surface will be

$$p = p_0 + p_1 = \text{const.} \tag{2}$$

There is, however, another condition to be fulfilled which is of equal importance. The free surface must move with the liquid and not independently of it. This (Ref. 3, p. 363) may be formulated as

$$\mathrm{D}p/\mathrm{D}t = 0 \text{ at the free surface.} \tag{3}$$

Retaining only first-order terms, this becomes

$$\frac{\partial p_1}{\partial t} + p_0{}'\mathbf{u}_1 \cdot \boldsymbol{\zeta} = 0 \quad \text{at } \chi = 0,$$

or because of the hydrostatic equation,

$$\frac{\partial p_1}{\partial t} = \rho_0 g \mathbf{u}_1 \cdot \boldsymbol{\zeta} \quad \text{at } \chi = 0. \tag{4}$$

Introducing the vertical displacement from Eq. (9–10), this is

$$\frac{\partial p_1}{\partial t} = \rho_0 g \frac{\partial h_1}{\partial t} \quad \text{at } \chi = 0,$$

which may be integrated to

$$p_1 = \rho_s g h_1 \quad \text{at} \quad \chi = 0. \tag{5}$$

References for Chapter II

1. V. Bjerknes, J. Bjerknes, H. Solberg and T. Bergeron, *Physikalische Hydrodynamik*, Springer, Berlin, 1933.
2. C. Eckart and H. G. Ferris, Equations of motion of the ocean and atmosphere, *Rev. Modern Phys.* **28**, 48–52 (1956).
3. H. Lamb, *Hydrodynamics*, Cambridge Univ. Press, 6th ed., 728 pp., 1932.
4. P. S. Laplace, *Mécanique Céleste*, 4 vols. (trans. by N. Bowditch), Boston, 1832.

CHAPTER III

STEADY MOTIONS

12. Introduction

In general, the solutions of the first-order equations will represent motions that change with time, but they have particular solutions that represent steady states. It is convenient to begin with a study of these steady solutions for which all time derivatives are zero. Any quantity, such as velocity or pressure, may still vary in space.

The question arises whether such solutions are at all relevant, for the temporal changes observed in the atmosphere, and also in the oceans, are striking and complex. To some extent, these temporal changes are over-emphasized by the sensitivity of the human body. A ten per cent change in atmospheric temperature or pressure will have very much larger effects on human metabolism. A less-biased consideration of these matters might rather emphasize the smallness of the departures from a steady state: these are almost all below the ten per cent level. The single exception is the wind velocity—but this is almost always a small fraction of the velocity of sound, and can be treated as small for this reason.

Mean pressure. If any atmospheric variable, such as wind or pressure, is averaged over a number of years, it becomes apparent that there is a geographically variable, but temporally constant, component (Fig. 4). This is conveniently called the steady or permanent component and, because of the linearity of the first-order equations, can be considered independently of the components that change with time.

The steady component of the atmospheric pressure may be further resolved into three components

$$p = p_0(\chi) + p_Z(\phi, \chi) + p_T(\lambda, \phi, \chi),$$

where χ is the altitude, ϕ is the latitude, and λ the longitude. The

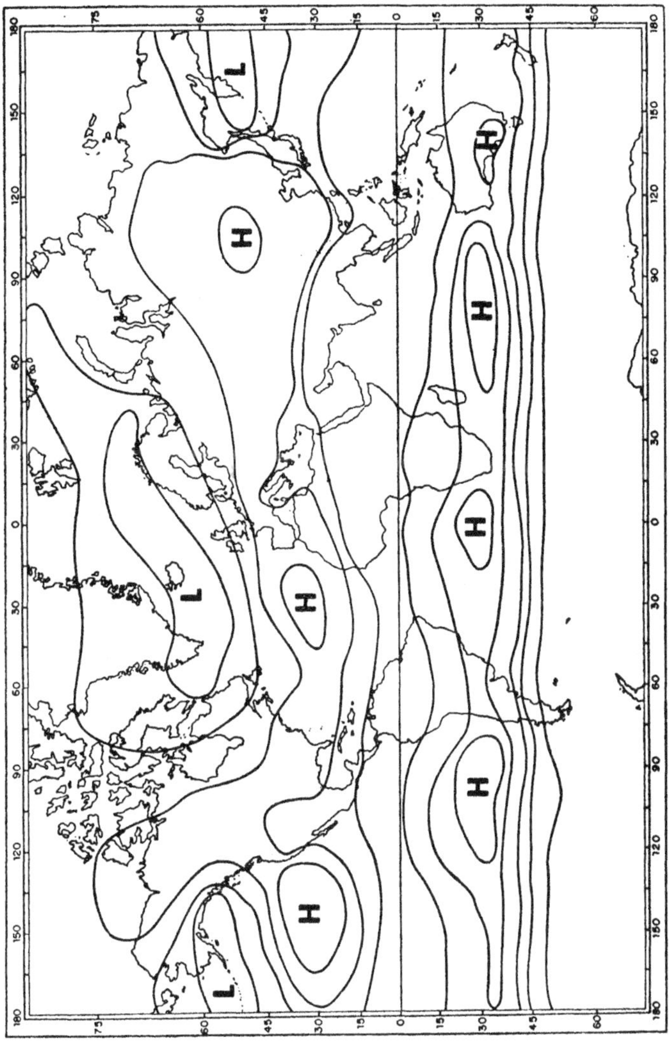

FIG. 4. Mean pressure of the atmosphere at sea level. Contour interval, 3 mb.

first of these components is barotropic and may be identified with the zero-order pressure previously introduced. The second depends only on latitude and altitude: it may be called the *zonal component*. The third depends on all three co-ordinates and may be called the *tesseral component*.

The zonal component. The sea-level values of p are shown on Fig. 4; by averaging these over longitude, the sum $p_0 + p_Z$ is obtained. For reasons that will appear later, it is convenient to define p_Z so that it vanishes at the equator: then it is found that

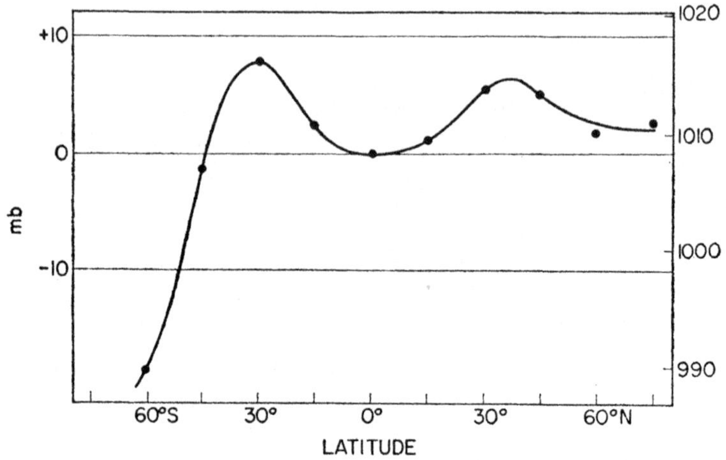

FIG. 5. Zonal component of the mean pressure of the atmosphere at sea level.

$p_0 = 1010$ mb at sea level. Figure 5 shows the sea-level values of the zonal component. It is seen to have a pronounced minimum at or near the equator, and to become negative at about the latitude of Cape Horn. Except in Antarctic latitudes, its numerical value does not exceed one per cent of p_0. In general, this small proportion holds at most altitudes, and is a justification for treating the zonal component as a perturbation of the barotropic, zero-order component $p_0(\chi)$.

While the general characteristics shown on Fig. 5 are doubtless correct, some allowance must be made for observational error. The observations are not well distributed, and there are few from the oceanic areas. Consequently some features (e.g. the curvature of the graph at the equator) may be quantitatively in error.

The tesseral component. Figure 6 shows the sea-level values of the tesseral component, obtained from Fig. 4 by subtracting $p_0 + p_Z$. Over the greater part of the earth's surface, its value does not exceed one-half per cent of p_0, again justifying its treatment as a perturbation. Along the equator, its values differ from zero by little, if any, more than the anticipated errors. At higher latitudes, its contours show various high and low pressure areas. It is thought that the general location of these is accurate, although their contours are certainly affected by the rather large errors in the data and its analysis.

The two continental highs, over North America and Asia, are obscured in many presentations not only by the large zonal gradient but also by large seasonal and other variations. The oceanic highs are more outstanding and are associated with the trade winds. The highs are arranged roughly along three circles of latitude, and are separated by lows at about the same latitudes. Most striking is the asymmetry of the northern and southern hemispheres, which is so great that it can scarcely be ascribed to observational error, and is most probably a reflection of the continental asymmetry.

Figures 4 to 6 give data only for sea level. Similar charts could be prepared for higher elevations. A striking phenomenon in the stratosphere is the high zonal wind at mid-latitudes, called the jet stream. It is associated with a relatively strong gradient of the zonal pressure. Because of this, it might be desirable to include the zonal component in the zero-order solution and not to treat it as a perturbation. This would be possible but would greatly complicate not only the following but also the preceding chapters. The possibility of this alternative method of calculation will not be explored in this book.

The oceans. Turning from the atmosphere to the oceans, these also exhibit steady components of velocity and pressure. The best known of these are the Gulf Stream and the Kuroshio–Oyoshio current systems. The analysis of the oceanic situation is less simple than that of the atmosphere. Consequently this chapter will be focused largely on the atmospheric phenomena described above.

The atmosphere and oceans as a heat engine. It is generally supposed that all motions in the atmosphere–ocean system are caused by the conversion of solar radiation into kinetic energy. This is sometimes expressed by saying that this system is a heat engine, but this analogy can be misleading. The usual definition of a heat

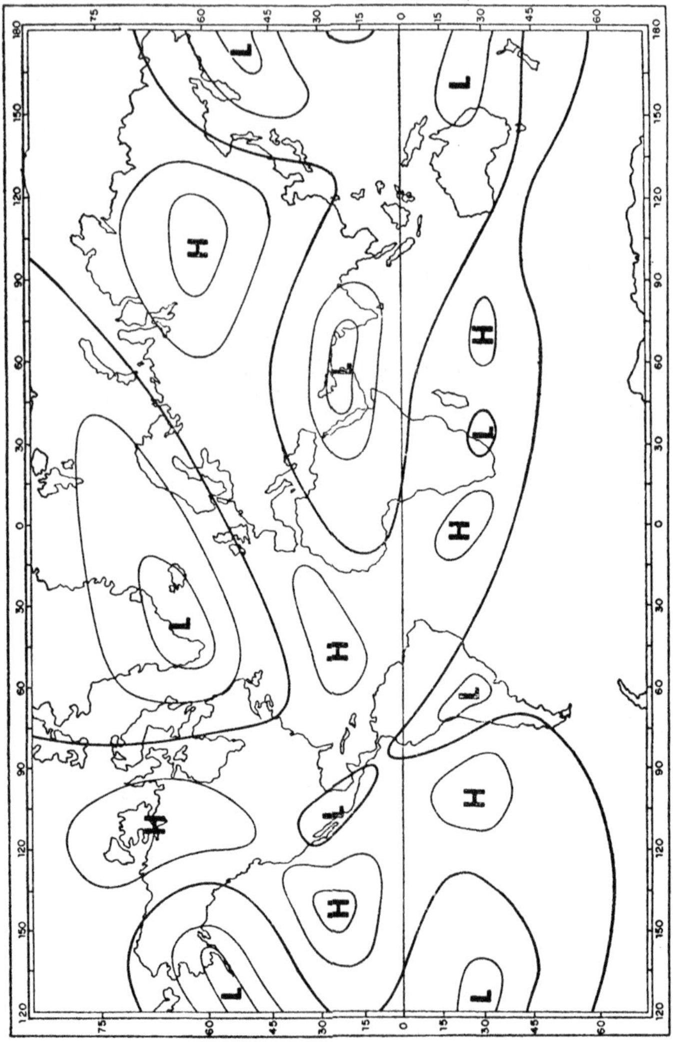

FIG. 6. Tesseral component of the mean pressure of the atmosphere at sea level.

engine implies a system that does work on its surroundings. If one neglects such minor incidents as erosion and the occasional destruction of trees and man-made structures, the atmosphere–ocean system seems to do no external work—on the average. There is a seasonal interchange of angular momentum between the solid earth and its fluid envelope, which is evidenced by a seasonal variation in the length of day; the amount of energy involved is small, and averages to zero over a period of years. It is therefore most reasonable to suppose that, to the first order, the atmosphere–ocean system does no external work, and differs from man-made heat engines in this important respect.

Every real heat engine also involves irreversible internal processes, that generate entropy. The most common are friction (or viscosity) and thermal conduction. A simple calculation shows that these are completely negligible for the large-scale motions of either ocean or atmosphere. Even in the turbulent boundary layers, the irreversible aspects of the phenomena are minor. It is therefore reasonable to suppose (again to first order) that the atmosphere–ocean system is a reversible one, in the thermodynamic sense of the term. This implies only that the system does not, to this approximation, generate entropy.

It does not imply the absence of those phenomena that are loosely described by the phrase "turbulent dissipation," and which are also irreversible, though in another sense of the term. Two kinds of non-thermodynamic irreversibility can be distinguished. The first is sometimes called "the spreading of wave-packets" and occurs whenever the velocity of simple waves in a medium is a function of frequency. The foundations for a treatment of these phenomena will be laid in subsequent chapters. The second results from the coupling between eigen solutions, induced by (1) the non-linearity of the basic equations and (2) the small-scale topography of the boundaries. This second kind of turbulent dissipation is entirely outside the scope of this book.

The above hypotheses, together, imply that $\mathbf{f}_1 = 0$; to the second and higher orders, there may be forces that are not included on the left side of the hydrodynamic equations. These hypotheses also imply something in regard to the net accession of heat, q_1. It has already been noted, in Section 5, that the heating of the atmosphere is primarily due to the balance between the solar radiation absorbed and the infra-red radiation emitted by the atmosphere. These processes are doubtless influenced by the motion of the atmosphere, but the coupling between the radiative and mechanical processes is apparently weak.

This coupling is undoubtedly a very complex process, which is not yet understood even in its broadest features. It is almost certainly one of the major causes (if not the only one) for the climatic instability evidenced by the geologic and anthropologic records. If one ignores this instability, however, the details of the process need not concern one.

Instead, it is possible to appeal to observation, and calculate from this, the result of the coupling. The observation is simple: ignoring climatic instability, the atmosphere does have a steady component to its motion. The question to be answered by the theory is then: what restrictions must be fulfilled by the steady component of the net accession of heat, q_1, in order that such a steady component of the atmospheric motions may exist?

It is to be assumed that the complex processes mentioned above conspire to cause the net accession of heat to fulfill these restrictions.

The equations. In view of the foregoing, the investigation of steady motions is certainly not irrelevant, and this chapter will be devoted to them. It will be based on Eqs. (9–2, 3, 4), as simplified by setting all time derivatives equal to zero. These equations then reduce to

$$\nabla p_1 + \rho_1 g \boldsymbol{\zeta} + \rho_0 \boldsymbol{\Omega} \times \mathbf{u}_1 = \rho_0 \mathbf{f}_1, \tag{1}$$

$$\nabla \cdot (\rho_0 \mathbf{u}_1) = 0, \tag{2}$$

$$\eta_0' \mathbf{u}_1 \cdot \boldsymbol{\zeta} = q_1/\theta_0. \tag{3}$$

These are the first-order equations of steady motion. If $\mathbf{f}_1 = 0$ and $q_1 = 0$, the motion is "free"; if $\mathbf{f}_1 = 0$ but $q_1 \neq 0$, the motion is "purely convective." Only these two cases will be treated below, although it would be possible to treat forced convection by means of hypotheses concerning the force $\mathbf{f}_1$.

13. Free steady motion, no rotation

The simplest problems are presented by free motion without Coriolis force. The equations to be solved are then

$$\nabla p_1 + \rho_1 g \boldsymbol{\zeta} = 0, \tag{1}$$

$$\nabla \cdot (\rho_0 \mathbf{u}_1) = 0, \tag{2}$$

$$\eta_0' \mathbf{u}_1 \cdot \boldsymbol{\zeta} = 0. \tag{3}$$

The first of these equations is independent of the velocity, and yields only the additive barotropic terms that have already been discussed in Section 10. They will be ignored in what follows.

Equation (13–3) factorizes into the alternatives $\eta_0' = 0$ and $\mathbf{u}_1 \cdot \boldsymbol{\zeta} = 0$. The former case can occur only if the stratification is isentropic, and will be considered in Section 28. The second case reduces Eqs. (2) and (3) to

$$\nabla \cdot \mathbf{u}_1 = 0, \tag{4}$$

$$\mathbf{u}_1 \cdot \boldsymbol{\zeta} = 0. \tag{5}$$

The motion is therefore purely horizontal, but will in general depend upon the altitude χ.

The general solution of Eqs. (13–4, 5) is

$$\begin{aligned} \mathbf{u}_1 &= \nabla \times (\chi\psi) \\ &= \nabla\chi \times \nabla\psi \\ &= \boldsymbol{\zeta} \times \nabla\psi, \end{aligned} \tag{6}$$

where ψ is the *stream function.* It is often convenient to use the notation

$$\mathbf{M} = \boldsymbol{\zeta} \times \nabla. \tag{7}$$

Plane level surfaces. In this case, $\boldsymbol{\zeta} = \mathbf{k}$, and $\chi = z$, so that

$$\mathbf{M} = -\mathbf{i}\frac{\partial}{\partial y} + \mathbf{j}\frac{\partial}{\partial x},$$

and if

$$\mathbf{u}_1 = \mathbf{i}u_x + \mathbf{j}u_y + \mathbf{k}u_z,$$

Eq. (13–6) becomes

$$u_x = -\frac{\partial\psi}{\partial y}, \qquad u_y = \frac{\partial\psi}{\partial x}, \qquad u_z = 0. \tag{8}$$

The equations of the streamlines are

$$\psi(x, y, z) = a, \qquad z = b, \tag{9}$$

a and b being constants. They are level plane curves, and no two intersect except at points of stagnation.

They must be tangent to any stationary boundary; and since they do not intersect, Eq. (13–9), for some values of a and b, must represent the contours of the boundary. In the simplest case of level boundaries, this imposes no condition upon the stream function. More generally, let the bottom boundary of the oceans or atmosphere have the equation

$$z = B(x, y). \tag{10}$$

Then if the stream function is analytic in z, it must have the form

$$\psi(x, y, z) = F[B(x, y), z] + [B(xy) - z]^n G(x, y), \tag{11}$$

F and G being undetermined, and $n > 0$. Near the boundary, the function F will dominate the motion. Outside the boundary layer, the function G will dominate; the thickness of the boundary layer will be determined jointly by the functions F, G, and the exponent n. Within the boundary layer, the motion has a simple description: if cylinders are erected on the bottom contours with vertical axes, the stream-lines will be the horizontal sections of these cylinders. More briefly, these cylinders are stream surfaces. Such motion has been called "bathystrophic".[(4)] Outside the boundary layer, the second term causes the motion to be less dependent upon the bottom contours.

Spherical level surfaces. If $\chi = r - r_s$ (see Section 5), the level surfaces are spheres, and it is convenient to use the Mercator co-ordinates of the Appendix. Then

$$\mathbf{M} = \frac{\cosh \mu}{r}\left(-\boldsymbol{\epsilon}\frac{\partial}{\partial \mu} + \boldsymbol{\nu}\frac{\partial}{\partial \lambda}\right),$$

$$\mathbf{u}_1 = \boldsymbol{\epsilon} u_\epsilon + \boldsymbol{\nu} u_\nu + \boldsymbol{\zeta} u_\zeta,$$

and Eq. (13–6) is

$$u_\epsilon = -\frac{\cosh \mu}{r}\frac{\partial \psi}{\partial \mu}, \quad u_\nu = \frac{\cosh \mu}{r}\frac{\partial \psi}{\partial \lambda}, \quad u_\zeta = 0. \tag{12}$$

The close analogy between Eqs. (13–8) and (13–12) is due to the use of Mercator co-ordinates. The factor $1/r$ may, if desired, be absorbed in the function ψ, but the factor $\cosh \mu$ cannot be so absorbed without spoiling the analogy.

In order that the velocity shall not become infinite at the poles, the derivatives of ψ with respect to μ and λ must vanish there. The remaining discussion of the stream function is identical with that for the plane level surfaces.

14. Second-order instability; the secular equation

It is not proposed to give an exhaustive discussion of second-order phenomena in this book, but some reference to them is needed to avert false conclusions. In general, let

$$\begin{aligned} \rho &= \rho_0 + \rho_1 + \rho_2 \cdots, \\ p &= p_0 + p_1 + p_2 \cdots, \\ \mathbf{u} &= \qquad \mathbf{u}_1 + \mathbf{u}_2 \cdots. \quad \text{etc.} \end{aligned}$$

In Section 7, ρ_0 and p_0 were supposed independent of time; in the last section, ρ_1, p_1 and $\mathbf{u}_1$ were determined so that they are also independent of time.

It should not be supposed that these two assumptions imply that ρ_2, p_2, $\mathbf{u}_2$ are also steady. In fact, it will be shown that this is a *third* assumption, and imposes certain restrictions upon ρ_1, p_1, $\mathbf{u}_1$, in addition to those imposed by the first two. It is therefore convenient to distinguish between zero-, first-, and second-order steadiness of the motion. If second-order steadiness is also imposed on free motion, the equations

$$v_0 \nabla p_2 + v_2 p_0' \boldsymbol{\zeta} = -\mathbf{u}_1 \cdot \nabla \mathbf{u}_1 - v_1 \nabla p_1, \tag{1}$$

$$v_0' \mathbf{u}_2 \cdot \boldsymbol{\zeta} - v_0 \nabla \cdot \mathbf{u}_2 = -\mathbf{u}_1 \cdot \nabla v_1 + v_1 \nabla \cdot \mathbf{u}_1, \tag{2}$$

$$\eta_0' \mathbf{u}_2 \cdot \boldsymbol{\zeta} = -\mathbf{u}_1 \cdot \nabla \eta_1, \tag{3}$$

are obtained from considerations analogous to those of Section 9.

It is again significant that only v_2 and p_2 appear in Eq. (14–1), and not $\mathbf{u}_2$. If the zero- and first-order quantities are treated as knowns, Eq. (14–1) contains just two scalar unknowns. Since a vector equation is equivalent to three scalar equations, there are thus more equations than unknowns. However, it has just been seen that $p_1(\chi)$ and $\psi(x, y, z)$ are arbitrary. These can thus be considered as two additional scalar unknowns, making four unknowns and three equations—a much more comfortable mathematical situation.

Secular equations. This situation is very often encountered in perturbation calculations, both with ordinary and partial differential equations. It may be described in these terms: the general solution of the first-order equations involves arbitrary elements (constants or functions of integration). The second-order equations frequently are redundant, unless the arbitrary first-order elements are considered to be unknowns. In such cases, it is possible to eliminate the second-order variables and obtain an equation (or several equations) in which only first-order quantities appear. Such equations are in one sense second-order equations, since they can only be derived by making a second-order calculation. In another sense, they are first-order, since they contain no second-order terms. They are often quadratic, or at least non-linear. For convenience, they may be called secular equations, although this is a slight extension of the accepted meaning of the term.

It will, for the remainder of this section, be supposed that the level surfaces are plane; using the notation of the last section,

$$\mathbf{u}_1 \cdot \nabla = -\frac{\partial \psi}{\partial y}\frac{\partial}{\partial x} + \frac{\partial \psi}{\partial x}\frac{\partial}{\partial y}, \tag{4}$$

and so Eq. (14–1) becomes

$$v_0 \frac{\partial p_2}{\partial x} = -\left(\frac{\partial \psi}{\partial y}\frac{\partial}{\partial x} - \frac{\partial \psi}{\partial x}\frac{\partial}{\partial y}\right)\frac{\partial \psi}{\partial y}, \tag{5}$$

$$v_0 \frac{\partial p_2}{\partial y} = +\left(\frac{\partial \psi}{\partial y}\frac{\partial}{\partial x} - \frac{\partial \psi}{\partial x}\frac{\partial}{\partial y}\right)\frac{\partial \psi}{\partial x}, \tag{6}$$

$$v_0 \frac{\partial p_2}{\partial z} - \rho_0 g v_2 = v_1 \rho_1 g, \tag{7}$$

use having been made of the relations $p_0' = -\rho_0 g$, $p_1' = -\rho_1 g$. It is at once seen that Eqs. (14–5) and (14–6) contain only p_2 and ψ, and may be treated independently of Eq. (14–7). Since v_0 depends only on z, p_2 can be eliminated by cross-differentiation with respect to x and y, yielding the secular equation

$$\frac{\partial}{\partial y}\left(\frac{\partial \psi}{\partial y}\frac{\partial^2 \psi}{\partial x\,\partial y} - \frac{\partial \psi}{\partial x}\frac{\partial^2 \psi}{\partial y^2}\right) + \frac{\partial}{\partial x}\left(\frac{\partial \psi}{\partial y}\frac{\partial^2 \psi}{\partial x^2} - \frac{\partial \psi}{\partial x}\frac{\partial^2 \psi}{\partial x\,\partial y}\right) = 0. \tag{8}$$

It is quadratic, but contains only one unknown—the first-order stream function. The assumption of second-order steadiness therefore imposes this further condition upon the first-order motion. Unless it is satisfied, the motion will, in fact, not be steady, but only approximately so; these changes are commonly called secular variations and are usually, in some sense, slow. It is therefore seen that motions which are steady to first-order may very often be unstable with respect to second-order effects.

Relation to the vorticity theorem. The secular equation can be given another form: after carrying out the indicated differentiations, Eq. (14–8) becomes

$$\left(\frac{\partial\psi}{\partial y}\frac{\partial}{\partial x} - \frac{\partial\psi}{\partial x}\frac{\partial}{\partial y}\right)\left(\frac{\partial^2\psi}{\partial x^2} + \frac{\partial^2\psi}{\partial y^2}\right) = 0. \tag{9}$$

Now Eq. (13–8) shows that

$$\frac{\partial^2\psi}{\partial x^2} + \frac{\partial^2\psi}{\partial y^2} = \frac{\partial u_y}{\partial x} - \frac{\partial u_x}{\partial y} = \varpi \tag{10}$$

is the vorticity of the flow, and that

$$-\left(\frac{\partial\psi}{\partial y}\frac{\partial}{\partial x} - \frac{\partial\psi}{\partial x}\frac{\partial}{\partial y}\right) = \mathbf{u}_1 \cdot \nabla = \frac{\mathrm{D}}{\mathrm{D}t}. \tag{11}$$

Hence Eq. (14–9) becomes

$$\frac{\mathrm{D}\varpi}{\mathrm{D}t} = 0, \tag{12}$$

and states that the vorticity of every particle of the fluid remains constant during its motion. This is a special and approximate form of general theorems proven by Helmholtz and Bjerknes (Ref. 2, Chapter IV; Ref. 5, Chapter VII).

It is important to note that this theorem is related to the secular equation and not to the first-order equations themselves.

A first integral of the secular equation. Equation (14–9) can be integrated at once to yield

$$\frac{\partial^2\psi}{\partial x^2} + \frac{\partial^2\psi}{\partial y^2} = \varpi(\psi, z) \tag{13}$$

the vorticity ϖ being an arbitrary differentiable function of its arguments. Because of the arbitrariness of ϖ, Eq. (14–13) imposes only weak restrictions on the stream function ψ. Numerous special solutions of the equation can be found, although its general solution cannot be written out. Numerical methods for its solution are easily devised.

Relation to the Bernoulli theorem. Equation (14–5) may be written

$$\frac{\partial p_2}{\partial x} = -\rho_0(z)\left[\frac{\partial \psi}{\partial y}\frac{\partial^2 \psi}{\partial x\,\partial y} - \frac{\partial \psi}{\partial x}\frac{\partial^2 \psi}{\partial x^2}\right]$$

$$= -\rho_0(z)\left[\frac{\partial \psi}{\partial y}\frac{\partial \psi^2}{\partial x\,\partial y} + \frac{\partial \psi}{\partial x}\frac{\partial^2 \psi}{\partial x^2} - \varpi(\psi, z)\frac{\partial \psi}{\partial x}\right];$$

Eq. (14–6) may be similarly transformed. The resulting equations can then be integrated at once to yield

$$p_2 = -\rho_0\left\{\frac{1}{2}\left(\frac{\partial \psi}{\partial x}\right)^2 + \frac{1}{2}\left(\frac{\partial \psi}{\partial y}\right)^2 - \int \varpi(\psi, z)\, d\psi\right\} \tag{14}$$

or

$$\tfrac{1}{2}\rho_0 \mathbf{u}_1^{\,2} + p_2 = \rho_0 \int \varpi(\psi, z)\, d\psi. \tag{15}$$

This equation determines p_2, once ψ and ϖ have been fixed; the determination of v_2 then follows at once from Eq. (14–7).

Equation (14–15) will be recognized as an approximate form of the Bernoulli theorem (Ref. 6, pp. 21–23); this is again not a consequence of the first-order equations, but of the second-order equations. The relation between the vorticity and the Bernoulli theorem is significant.

15. Free steady motion with rotation

If the Coriolis force is included, the first-order equations of free steady motion become

$$\nabla p_1 + \rho_1 g \boldsymbol{\zeta} + \rho_0 \boldsymbol{\Omega} \times \mathbf{u}_1 = 0, \tag{1}$$

$$\nabla \cdot (\rho_0 \mathbf{u}_1) = 0, \tag{2}$$

$$\eta_0' \mathbf{u}_1 \cdot \boldsymbol{\zeta} = 0. \tag{3}$$

The geostrophic equation. As before it will be supposed that $\eta_0' > 0$, $\mathbf{u}_1 \cdot \boldsymbol{\zeta} = 0$. In this case, Eq. (15–1) involves $\mathbf{u}_1$ as well as p_1 and ρ_1; the second of these variables can be eliminated by multiplying each term vectorially with $\boldsymbol{\zeta}$:

$$\boldsymbol{\zeta} \times \nabla p_1 + \rho_0 \boldsymbol{\zeta} \times (\boldsymbol{\Omega} \times \mathbf{u}_1) = 0 \tag{4}$$

The last term on the left is

$$\boldsymbol{\zeta} \times (\boldsymbol{\Omega} \times \mathbf{u}_1) = \boldsymbol{\Omega}\boldsymbol{\zeta} \cdot \mathbf{u}_1 - \boldsymbol{\zeta} \cdot \boldsymbol{\Omega}\mathbf{u}_1, \tag{5}$$

which can be simplified because $\boldsymbol{\zeta} \cdot \mathbf{u}_1 = 0$. This reduces Eq. (15–4) to

$$\mathbf{u}_1 = \mathbf{M}p_1/\rho_0\boldsymbol{\Omega} \cdot \boldsymbol{\zeta}. \tag{6}$$

This is the well known geostrophic equation, and enables the calculation of the velocity once p_1 is known. It implies that the stream lines are the contours of the isobaric surfaces, and that the direction of motion along them is such that, in the northern hemisphere, the higher pressure is to the right, if one imagines walking on the isobar in the direction of flow.

Relation between density and pressure. Analogously, scalar multiplication with $\boldsymbol{\Omega}$ eliminates $\mathbf{u}_1$ from Eq. (15–1) and leads to the formula

$$\rho_1 g = -\boldsymbol{\Omega} \cdot \nabla p_1/\boldsymbol{\Omega} \cdot \boldsymbol{\zeta}, \tag{7}$$

which enables the calculation of ρ_1 from p_1.

Planetary vorticity. It remains to consider Eq. (15–2); substitution from the geostrophic equation, Eq. (15–6), converts it into

$$\nabla \cdot \{(\boldsymbol{\zeta} \times \nabla p_1)/\boldsymbol{\Omega} \cdot \boldsymbol{\zeta}\} = 0, \tag{8}$$

which is an equation in p_1 only. This equation can be much simplified. Note that, as in Eq. (13-6),

$$\boldsymbol{\zeta} \times \nabla p_1 = \nabla \chi \times \nabla p_1 = \nabla \times (\chi p_1)$$

and that the divergence of a curl vanishes. The numerator in the bracket of Eq. (15–8) can thus be treated as a constant in carrying out the indicated differentiation and, therefore

$$(\boldsymbol{\zeta} \times \nabla p_1) \cdot \nabla(\boldsymbol{\Omega} \cdot \boldsymbol{\zeta}) = 0.$$

Finally, define

$$\begin{aligned} \boldsymbol{\beta} &= -\boldsymbol{\zeta} \times \nabla(\boldsymbol{\Omega} \cdot \boldsymbol{\zeta}) \\ &= \nabla \times (\boldsymbol{\Omega} \cdot \boldsymbol{\zeta}\boldsymbol{\zeta}) \\ &= -\mathbf{M}(\boldsymbol{\Omega} \cdot \boldsymbol{\zeta}), \end{aligned} \tag{9}$$

whereupon Eq. (15–8) may be written

$$\boldsymbol{\beta} \cdot \nabla p_1 = 0. \tag{10}$$

The vector $\boldsymbol{\beta}$ has been called the planetary vorticity, and the last equation shows that if $\boldsymbol{\beta}$ does not vanish, the pressure gradient must be perpendicular to the planetary vorticity.

Plane level surfaces. If the level surfaces are planes, the coordinate system can be chosen so that $\mathbf{\Omega}$ lies in the (y, z) plane. Then

$$\boldsymbol{\zeta} = \mathbf{k},$$

$$\mathbf{\Omega} = \Omega_H \mathbf{j} + \Omega_V \mathbf{k},$$

and Eq. (15–6) becomes

$$\mathbf{u}_x = -\frac{1}{\rho_0 \Omega_V} \frac{\partial p_1}{\partial y}, \qquad \mathbf{u}_y = +\frac{1}{\rho_0 \Omega_V} \frac{\partial p_1}{\partial x}. \tag{11}$$

It is noteworthy that only the vertical component of the angular velocity, Ω_V, enters this equation; in later sections, it will be found that analogous equations depend on the horizontal component as well.

Equation (15–7) becomes

$$\rho_1 g = -\frac{\partial p_1}{\partial z} - \frac{\Omega_H}{\Omega_V} \frac{\partial p_1}{\partial y}, \tag{12}$$

which differs from the hydrostatic equation only because of the last term. Now, in both the atmosphere and oceans

$$\left| \frac{\partial p_1}{\partial y} \right| \ll \left| \frac{\partial p_1}{\partial z} \right|,$$

so that this term may usually be neglected.

Since $\boldsymbol{\zeta} \cdot \mathbf{\Omega} = \Omega_V =$ constant, the planetary vorticity vanishes in this case, and Eq. (15–10) imposes no restriction on p_1. In this case, also, the stream function is

$$\psi = p_1/\rho_0 \Omega_V;$$

the secular equation can therefore be expected to impose restrictions on the pressure, if second-order steadiness is required.

Spherical level surfaces. The Coriolis vector is now

$$\mathbf{\Omega} = \Omega(\boldsymbol{\nu} \operatorname{sech} \mu + \boldsymbol{\zeta} \tanh \mu) \tag{13}$$

(cf. Appendix, Eq. (9)) so that Eq. (15–6) becomes

$$\mathbf{u}_1 = \frac{\cosh \mu \operatorname{ctnh} \mu}{\rho_0 \Omega r} \left(-\boldsymbol{\epsilon} \frac{\partial p_1}{\partial \mu} + \boldsymbol{\nu} \frac{\partial p_1}{\partial \lambda} \right). \tag{14}$$

The general analogy to Eq. (15–11) will be evident. Equation (10) of the Appendix yields

$$\mathbf{\Omega} \cdot \nabla = \frac{\Omega}{r} \frac{\partial}{\partial \mu} + \Omega \tanh \mu \frac{\partial}{\partial r}$$

so that Eq. (15–7) becomes

$$\rho_1 g = -\frac{\partial p_1}{\partial r} - \frac{\operatorname{ctnh} \mu}{r} \frac{\partial p_1}{\partial \mu}, \tag{15}$$

Since $\mathbf{u}_1$ and p_1 must be finite everywhere, it follows that $\partial p_1/\partial \lambda$ and $\partial p_1/\partial \mu$ must vanish both at the poles ($\mu = \neq\infty$) and at the equator ($\mu = 0$). In general, the second term on the right of Eq. (15–15) will again be small compared to the first.

In this case, $\mathbf{\Omega} \cdot \boldsymbol{\zeta} = \Omega \tanh \mu$ is not constant, and the planetary vorticity is

$$\boldsymbol{\beta} = \mathbf{M\Omega} \cdot \boldsymbol{\zeta} = \boldsymbol{\epsilon}\Omega \operatorname{sech} \mu/r. \tag{16}$$

Consequently, Eq. (15–9) shows that p_1 must be independent of longitude, and $\partial p_1/\partial \lambda$ must vanish identically. Equation (15–14) then shows that the northerly component of $\mathbf{u}_1$ vanishes, and the motion is zonal.

The curvature of the level surfaces combined with rotation thus restricts the variety of steady motions very decidedly, and the question arises: what happens to those motions that would be steady if the level surfaces were plane? In Chapter XVII it will be shown that these motions are not steady but slowly variable. The time variations depend on the angular velocity Ω, and vanish when there is no rotation, or when there is no curvature of the level surfaces.

Thus, the rotation of a spherical earth unstabilizes all except zonal geostrophic motions to *first* order; in the absence of rotation, these are unstable, at most, to second order. This unstabilizing effect of rotation appears to have been overlooked in the past.

16. Pure convection, no rotation

In pure convection, the resultant force, $\mathbf{f}_1$, vanishes everywhere, but the net (or resultant) accession of heat, q_1, does not. The vanishing of the mechanical force implies that there is no dissipation, no conversion of kinetic energy into heat. The fluid does work only against the force of gravity. This raises two questions. The one is purely theoretical: can steady motions occur under these conditions?

Assuming that this has an affirmative answer, the second and more difficult question is: do these motions have any relevance to the actual steady movement of the atmosphere or ocean?

The first question is not trivial; consider the case of a closed streamline. A parcel of fluid that moves in this streamline will undergo periodic changes in volume and pressure, much as does the working fluid in a Carnot engine. However, since the force of gravity is conservative, the total work done by (or on) this parcel of fluid, during one cycle, will vanish. If it is considered as a heat engine, the efficiency of this engine must therefore be zero. A simple application of Carnot's formula shows that this can be the case only if as much heat is removed from the parcel at a given temperature as is added to it at that same temperature. In other words, steady, purely convective motion is possible only for certain distributions of the heat sources and sinks. The problem is to determine these special distributions.

Neglecting rotation, the equations to be discussed are

$$\nabla p_1 + \rho_1 g \boldsymbol{\zeta} = 0, \tag{1}$$

$$\nabla \cdot (\rho_0 \mathbf{u}_1) = 0, \tag{2}$$

$$\mathbf{u}_1 \cdot \boldsymbol{\zeta} = q_1/\eta_0'\theta_0 = w. \tag{3}$$

As remarked in Section 5, q_1, as well as η_0' and θ_0, are to be treated as known data; the same will therefore be true of w, which is the vertical component of the velocity.

The vertical velocity. To obtain some estimate of orders of magnitude, one may first consider an isothermal atmosphere for which $\eta_0'\theta_0 = g$ (Section 8). A vertical velocity of 1 cm sec^{-1} therefore requires a net accession of 980 erg g^{-1} sec^{-1}; if the density of the air is 10^{-3} g cm^{-3}, this is about 1 erg cm^{-3} sec^{-1} or 0·75 cal/cm^3 per year. Now, the solar constant is $\sigma = 2$ cal cm^{-2} min^{-1}, so that the Earth intercepts solar energy at the rate $\pi r_s^2 \sigma$. Since the spherical surface is $4\pi r_s^2$, the mean insulation is $\sigma/4 = 0{\cdot}5$ cal cm^{-3} min^{-1} or $0{\cdot}26 \times 10^6$ cal cm^{-2} yr^{-1}. If this were all absorbed in a column of air 10 km high, the gross accession of heat would be 0·26 cal cm^{-3} yr^{-1}; and because of transparency and back radiation, the net accession will certainly be less than this. Consequently, one may conclude that, in an isothermal atmosphere, the vertical velocity, w, cannot exceed 1 cm sec^{-1} and is probably less. Because of latent heat effects, local values may be larger, but the limit just found will apply to large-scale phenomena.

More generally, (Section 7),

$$\eta_0'\theta_0 = C_{p0}(\theta_0' + \theta_A'),$$

so that

$$w = q_1/[C_{p0}(\theta_0' + \theta_A')]. \tag{4}$$

Now, if the heating occurred at constant pressure and without motion, the temperature would increase at the rate

$$\tau = q_1/C_{p0}, \tag{5}$$

and hence Eq. (16–4) may be written

$$w = \tau/(\theta_0' + \theta_A'). \tag{6}$$

This equation may be interpreted as follows: a parcel of fluid is in hydrostatic equilibrium in the temperature gradient θ_0'. If it is heated, it will become lighter, and must rise; if the heating occurred at constant pressure and if it always remained in hydrostatic equilibrium, it would rise with the velocity τ/θ_0'. However, as it rises, it encounters a reduced pressure, which causes an additional expansion and consequent cooling because of the work done by it on its surroundings. To remain in hydrostatic equilibrium, it must therefore rise with a velocity w, that is less than τ/θ_0'. The term θ_A' in the denominator of Eq. (16–6) takes quantitative account of this reduction in the rate of rise because of the adiabatic cooling.

Returning to estimates of order of magnitude: Eq. (16–6) may also be written

$$w = w_{\text{isotherm.}}/(1 + \theta_0'/\theta_A'), \tag{7}$$

$w_{\text{isotherm.}}$ being the velocity if the stratification were isothermal. Hence, if $\theta_0'/\theta_A' = 1$, a velocity of 1 cm sec^{-1} would be produced by 1·5 cal cm^{-3} yr^{-1} etc. However, if $\theta_0'/\theta_A' = -1$, then $w = \infty$. This indicates that negative temperature gradients tend toward instability and, in fact, that when $\theta_0' \leqslant \theta_A'$, the calculations of this chapter are certainly invalid.

Boundary conditions. It follows from Eq. (16–3) that the net accession of heat must be zero wherever the vertical velocity vanishes. This occurs at all level stationary boundaries; if w does not vanish at such boundaries, the fluid will become warmer (or colder) there, and the motion will not be steady.

If the stationary boundary is not level, w need not vanish there. This is a fundamental distinction between level and inclined boundaries and requires further consideration. In any case, the streamlines must lie in the boundary surface; let P be a point at which the boundary is inclined at an angle α to the horizontal, and the streamline through P, at an angle β; then

$$-\alpha \leqslant \beta \leqslant +\alpha.$$

Let u be the magnitude of the velocity at P, so that the vertical component is

$$w = u \sin \beta.$$

It follows at once that

$$u > w/\sin \alpha \tag{8}$$

For given w, the minimal velocity at a stationary boundary is greater, the more nearly level that boundary is. If the velocity at a truly level boundary is to be both finite and steady, the net accession of heat must vanish there.

In many laboratory experiments on convection, considerable care is exercised to insure that the fluid has a level boundary and that the net accession of heat at the boundary does not vanish. For example, if a shallow pan of fluid is heated from below, this will be the case. Under these conditions, the convection will usually not be steady, but turbulent. Reasoning by analogy to such turbulent convective experiments, it is often supposed that all geophysical convection must be turbulent. This argument is not sound; and even if the conclusion should be correct, it is still important to find the conditions under which steady convection could occur. If it can then be demonstrated that these conditions do not arise in nature, it will become possible to assign a cause to the unsteady motions that are observed.

Consequences of the conservation of matter. A somewhat different condition is imposed upon q_1 if the fluid is contained in a basin with closed contours. Then, from Eq. (16–2) and the fact that no fluid escapes through the solid boundary of the basin, it follows that

$$\iint \mathbf{u}_1 \cdot \boldsymbol{\zeta} \, dS = 0,$$

the surface integral being extended over that portion of the level surface inside the contour at any level χ. Since η_0' and θ_0 depend only on χ, Eq. (16–3) converts this into

$$\iint q_1 \, dS = 0. \tag{9}$$

Under these conditions, therefore, steady convection can occur only if the total accession of heat is zero at every level beneath the sill of the basin.

The same argument shows that the total accession of heat must vanish at every level in a planetary atmosphere with spherical level surfaces. However, if the idealization of plane level surfaces extending to infinity is adopted, so that there are no barriers to the horizontal flow of the fluid, there is no analogue to Eq. (16–9), and the total accession of heat at any level may be finite. Then it is merely necessary that some streamlines extend to infinity at that level.

Since the zero-order temperature is a function only of χ, Eq. (16–9) may be rephrased: the total accession of heat at any temperature is, to first order, zero. Consequently, every heat source at a given temperature is balanced by an equal heat sink at that same temperature. Under these conditions, the Carnot formula yields zero as the efficiency of the heat engine, in conformity with the considerations given above.

The theoretical question has now been answered: the atmosphere (or ocean) can move with zero thermal efficiency and without departing from a steady state, provided the heat sources are distributed in accordance with Eq. (16–9). The geophysical question remains: are they so distributed, and if so, what mechanisms bring this about?

Explicit solution of the equations. Returning to the solution of the equations of motion, they may be written

$$\begin{aligned} \nabla \cdot \mathbf{u}_1 &= -\rho_0' w/\rho_0, \\ \boldsymbol{\zeta} \cdot \mathbf{u} &= w. \end{aligned} \tag{10}$$

Their solution is

$$\mathbf{u}_1 = \mathbf{M}\psi - \nabla V \tag{11}$$

where the velocity potential, V, is determined by

$$\begin{aligned} \nabla^2 V &= -\rho_0' w/\rho_0, \\ \boldsymbol{\zeta} \cdot \nabla V &= -w, \end{aligned} \tag{12}$$

while the stream function, ψ, is arbitrary.

Plane level surfaces. In this case, Eqs. (16–12) reduce to

$$\frac{\partial^2 V}{\partial x^2} + \frac{\partial^2 V}{\partial y^2} = \Delta, \tag{13}$$

where the vertical divergence is given by

$$\Delta = \frac{1}{\rho_0}\frac{\partial}{\partial z}(\rho_0 w) \tag{14}$$

and

$$\begin{aligned} u_x &= -\frac{\partial \psi}{\partial y} - \frac{\partial V}{\partial x}, \\ u_y &= +\frac{\partial \psi}{\partial x} - \frac{\partial V}{\partial y}, \\ u_z &= w. \end{aligned} \tag{15}$$

Any stationary boundaries will, of course, impose additional conditions on ψ and V.

Spherical level surfaces. In terms of Mercator co-ordinates, the equations become

$$\frac{\partial^2 V}{\partial \lambda^2} + \frac{\partial^2 V}{\partial \mu^2} = r^2(\operatorname{sech}^2 \mu)\,\Delta, \tag{16}$$

where the vertical divergence is

$$\Delta = \frac{1}{\rho_0 r^2}\frac{\partial}{\partial r}(\rho_0 r^2 w), \tag{17}$$

while

$$\begin{aligned} u_\epsilon &= \frac{\cosh \mu}{r}\left(-\frac{\partial \psi}{\partial \mu} - \frac{\partial V}{\partial \lambda}\right), \\ u_\nu &= \frac{\cosh \mu}{r}\left(+\frac{\partial \psi}{\partial \lambda} - \frac{\partial V}{\partial \mu}\right), \\ u_\zeta &= w. \end{aligned} \tag{18}$$

Again, the use of Mercator co-ordinates accentuates the analogy between the formulae for plane and spherical level surfaces.

17. Pure convection, with rotation

If the suppositions of the last section are modified only by including rotation, the equations to be solved are

$$\nabla p_1 + \rho_1 g \boldsymbol{\zeta} + \rho_0 \boldsymbol{\Omega} \times \mathbf{u}_1 = 0, \tag{1}$$

$$\nabla \cdot (\rho_0 \mathbf{u}_1) = 0, \tag{2}$$

$$\mathbf{u}_1 \cdot \boldsymbol{\zeta} = q_1/\eta_0'\theta_0 = w. \tag{3}$$

Since the last two equations are the same as before, the net accession of heat is subject to all the restrictions discussed in the previous section.

Thermobaric motion. As in Section 15, Eq. (17–1) yields

$$\boldsymbol{\zeta} \times \nabla p_1 + \rho_0 \boldsymbol{\zeta} \times (\boldsymbol{\Omega} \times \mathbf{u}_1) = 0,$$

and the identity, Eq. (15–5), reduces this to

$$\mathbf{u}_1 = (\mathbf{M} p_1/\rho_0 \boldsymbol{\Omega} \cdot \boldsymbol{\zeta}) + (\boldsymbol{\Omega} w/\boldsymbol{\Omega} \cdot \boldsymbol{\zeta}), \tag{4}$$

which is analogous to Eq. (15–6) and determines the velocity in terms of p_1 and w. In addition to the geostrophic motion, the velocity now has a component

$$\boldsymbol{\Omega} w/\boldsymbol{\Omega} \cdot \boldsymbol{\zeta}$$

parallel to the axis of rotation; this will be called the thermobaric motion. It is noteworthy that this thermobaric component is independent of the magnitude of the angular velocity, but depends upon the angle between its axis and the vertical.

A similar situation is encountered in acoustics where certain phenomena are caused by viscosity but are independent of the magnitude of the coefficient of viscosity.[3.7] Such phenomena occur only as steady components and may be called autostatic. Roughly speaking, they are the result of two balanced forces whose magnitudes are proportional to the same quantity (in this case, Ω). This quantity then cancels out of the equilibrium or steady state equations. In all autostatic phenomena, however, the approach to the steady state depends strongly on the parameter concerned.

Density and pressure. Equation (17–1) also yields the equation

$$\rho_1 g = -\boldsymbol{\Omega} \cdot \nabla p_1/\boldsymbol{\Omega} \cdot \boldsymbol{\zeta}, \tag{5}$$

which is identical with Eq. (15–7) and requires no further discussion. Substitution of Eq. (17–4) into Eq. (17–2) yields

$$\nabla \cdot (\boldsymbol{\zeta} \times \nabla p_1/\boldsymbol{\Omega} \cdot \boldsymbol{\zeta}) = -\delta \tag{6}$$

where δ is the divergence of the thermobaric flow of mass:

$$\delta = \boldsymbol{\Omega} \cdot \nabla(\rho_0 w/\boldsymbol{\Omega} \cdot \boldsymbol{\zeta}). \tag{7}$$

As in Section 15, Eq. (17–6) may be rewritten in terms of the planetary vorticity:

$$\boldsymbol{\beta} \cdot \nabla p_1 = (\boldsymbol{\Omega} \cdot \boldsymbol{\zeta})^2 \, \delta, \tag{8}$$

and is, again, formally an equation for the determination of p_1, in terms of δ.

Equation (17–8) is a more or less direct consequence of the equation

$$\nabla \cdot (\rho_0 \mathbf{u}_1) = 0$$

and therefore expresses the conservation of matter. Since it also contains the planetary vorticity, Eq. (17–8) may be named "the planetary divergence equation."

Plane level surfaces. In this case, the planetary vorticity, β, vanishes, and Eq. (17-8) reduces to $\delta = 0$ and restricts δ rather than p_1. Using the same notation as in Section 15, this implies

$$\left(\Omega_H \frac{\partial}{\partial y} + \Omega_V \frac{\partial}{\partial z}\right)(\rho_0 w) = 0. \tag{9}$$

The general solution of this equation is

$$\rho_0 w = F(x, y - \Omega_H z/\Omega_V), \tag{10}$$

F being an arbitrary differentiable function of its arguments.

If the fluid has a horizontal boundary, say at $z = 0$, then w must vanish there. Hence $F(x, y) = 0$ for all x and y, and w must vanish identically. There can be no steady convection in this very special case.

If the boundaries of the fluid are not level, the function F may be specified arbitrarily. The pressure, p_1, can also be specified arbitrarily except for the condition that the streamlines lie in the boundary.

Spherical level surfaces. In this case, the planetary vorticity does not vanish, and the planetary divergence equation, Eq. (17–8), reduces to

$$\frac{\partial p_1}{\partial \lambda} = \Omega r \tanh^2 \mu \left(\frac{\partial}{\partial \mu} + r \tanh \mu \frac{\partial}{\partial r} \right) (\rho_0 w \operatorname{ctnh} \mu), \tag{11}$$

as follows from Eq. (15–16). In general, it determines p_1 in terms of w; conversely, if p_1 is known, w can be found. The further discussion of this equation will be deferred to the next section.

18. Hadley's hypothesis of zonal heating

Since it is beyond the scope of this book to develop a theory of the net accession of heat, any examples of pure convection must be based on some hypothesis about q_1. In the case of a rotating spherical planet receiving radiant energy from a sun, q_1 will presumably be subject to large daily and seasonal variations and, in addition, will depend on both latitude and longitude. If the time variations are averaged out, a steady component of q_1 will be obtained: only this steady component is to be considered in this chapter.

Because of rotation, it is not entirely unreasonable to suppose that this q_1 is independent of longitude and depends only on latitude and altitude:

$$q_1 = q_1(\chi, \phi).$$

This will be called the hypothesis of zonal heating. It was first advanced by George Hadley in 1735.[1,5] It should be noted, however, that there are *a priori* reasons for doubting this hypothesis in the case of the Earth: land and sea are not distributed zonally, and their varying characteristics may well introduce a longitude dependence into q_1.

Because of its simplicity, Hadley's hypothesis of zonal heating has had considerable influence on meteorological thought, but it is now in some disrepute.[8] It will, none the, less be examined in detail because of its historical interest. It will suffice to consider only the isothermal atmosphere (Section 8).

No rotation. If rotation be neglected, the Eqs. (16–15, 17, 18) are applicable. Suppose that

$$w = (w_m \chi / h) \exp (1 - \chi / h) \cos 3\phi, \tag{1}$$

so that there is heating in the equatorial zone $|\phi| < 30°$, and cooling elsewhere; the total accession of heat at any altitude vanishes, in accordance with Eq. (16–9). The maximum upward velocity occurs at $\chi = h$, $\phi = 0$, and is w_m. If h is a small fraction of the radius of the solid planet, r_s, the vertical divergence is, to a sufficient approximation,

$$\Delta = \frac{1}{\rho_0} \frac{\partial}{\partial \chi} (\rho_0 w), \tag{2}$$

and Eq. (16–16) may be written

$$\frac{\partial^2 V}{\partial \lambda^2} + \frac{\partial^2 V}{\partial \mu^2} = r_s^2 \operatorname{sech}^2 \mu \, \Delta. \tag{3}$$

Since, for an isothermal atmosphere

$$\rho_0 = \rho_s \exp(-\chi/H), \tag{4}$$

where $H = R\theta_0/g$, Eqs. (18–1, 2) yield

$$\Delta = (w_m/h)(1 - \chi/h_1) \exp(1 - \chi/h) \cos 3\phi, \tag{5}$$

where

$$1/h_1 = 1/h + 1/H.$$

The vertical divergence thus vanishes at an altitude, h_1, that is lower than the altitude of maximum vertical velocity.

Equation (18–3) has a zonally symmetric solution for which the eastward component of velocity, u_ϵ, vanishes, while the northward component is, to a sufficient approximation

$$\begin{aligned} u_\nu &= -(\cosh \mu/r)(\partial V/\partial \mu) \\ &= -(w_m r_s/h)(1 - \chi/h_1) \exp(1 - \chi/h) \cdot \sin \phi \cos^2 \phi. \end{aligned} \tag{6}$$

The streamlines of this flow are shown on Fig. 7 for the case $H \gg h$; it is seen to consist of two ring vortices (the Hadley cells) whose axes are at $\chi = h_1$, $\phi = \pm 30°$. At low altitudes, the motion is equatorward, while for $\chi > h_1$, it is poleward. For other values of h/H, this is qualitatively unaltered.

At ground level, $\chi = 0$, the equator-ward velocity is

$$v = (w_m \, e \, r_s/h) \sin \phi \cos^2 \phi,$$

which has a maximum value of

$$v_m = w_m(r_s/h)(2 \, e \, \sqrt{3}/9) \tag{7}$$

at the latitudes where

$$\tan^2 \phi = 1/2,$$

or $\phi = \pm 35^\circ \, 16'$. It is noteworthy that this maximum ground wind occurs inside the zone of net cooling.

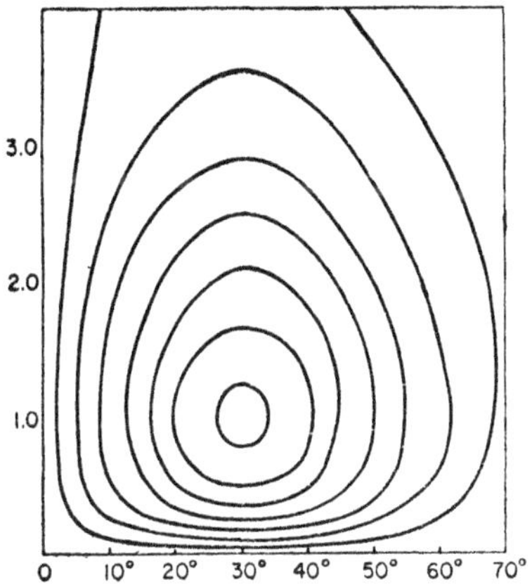

FIG. 7. The Hadley cell.

In the case of the Earth, $r_s = 6500$ km, while h is perhaps a few kilometers at most. It may therefore be concluded that $v_m/w_m \sim 10^3$. Since it has been seen that w may be as great as 1 cm/sec, convection might produce ground winds as great as 10 m/sec, which is considerably in excess of the observed steady component of the ground winds. Thus far, one must therefore conclude that Hadley's hypothesis is tenable: zonal heating by the sun could produce appreciable equatorward winds on the Earth.

Instability of the Hadley vortices. It remains to consider the stability of the motion just described. On a stationary Earth, it might be rendered unstable by *second*-order effects. This will not be investigated; instead it will be shown that rotation renders them unstable to *first* order. This is a direct consequence of the planetary divergence equation Eq. (17–11).

This equation is a necessary condition for steady motion on a rotating spherical Earth. If the heating is zonal, w, and therefore the right-hand side of this equation, must be independent of the longitude λ. The same must be true of $\partial p_1/\partial\lambda$, so that p_1 is, at most, a linear function of λ. Continuity of pressure then requires that it be independent of λ, i.e. that the right-hand side of Eq. (17–11) vanishes. This is not the case for the function considered above, nor can it be true for any physically reasonable distribution of vertical velocity.

This conclusion rests on two hypotheses:

(1) The hypothesis of pure convection,

(2) The hypothesis of zonal heating.

They cannot both be tenable since together they would imply a constantly accelerating equator-ward flow, and this is not the case.

19. Analysis of the Earth's permanent pressure field

In view of the *a priori* doubt concerning the validity of the hypothesis of zonal heating, it will not be considered further. The hypothesis of pure convection will now be examined in terms of the observed steady pressure field of the earth, as partially described in Section 12.

The zonal component of pressure. The planetary divergence equation affords the possibility of calculating the vertical velocity, w, from the observed departure of the pressure field from a barotropic. Since this equation involves only $\partial p_1/\partial\lambda$, the vertical velocity will be independent of the zonal pressure component. This is additional justification for the separation of this component from the tesseral, as in Section 12.

However, the zonal component will be associated with free geostrophic motion along circles of latitude, given by the equation

$$u_\epsilon = -\frac{1}{\rho_0 r\Omega\sin\phi}\frac{\partial p_Z}{\partial\phi}. \tag{1}$$

This velocity will be infinite at the equator ($\phi = 0$) unless $\partial p_Z/\partial\phi$ vanishes there. Figure 5 shows that this derivative does, in fact, vanish at or near the equator. The empirical material is not adequate to establish the position of this minimum more closely than $\pm 5°$, but is not inconsistent with its location at the equator. Assuming this, the dependence of the sea level velocity, u_ϵ, on latitude may be calculated from Fig. 5, and is shown on Fig. 8. The tropical zone is characterized by strong westward motion, while the temperate zones are the site of eastward winds. There is a marked asymmetry of this component of the wind which reaches nearly 9 m/sec at 45° south, but is scarcely more than 1 m/sec at 45° north.

A similar analysis could be made at other levels in the atmosphere. It is significant to note that the stratosphere is the site of very high velocities—the jet stream. Because of the factor ρ_0 in the denominator of Eq. (19–1), these unusually high velocities do not imply unusual pressure gradients.

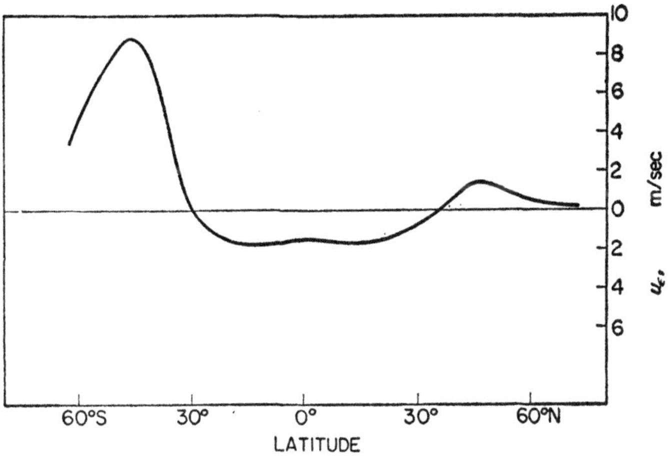

FIG. 8. The free geostrophic wind at sea level.

The tesseral component of pressure. According to Eqs. (17–4, 11) which are

$$\mathbf{u}_1 = (\mathbf{M}p_1/\rho_0\mathbf{\Omega} \cdot \boldsymbol{\zeta}) + (\mathbf{\Omega}w/\mathbf{\Omega} \cdot \boldsymbol{\zeta}), \tag{2}$$

$$\frac{\partial p_1}{\partial \lambda} = \Omega r \tanh^2 \mu \left(\frac{\partial}{\partial \mu} + r \tanh \mu \frac{\partial}{\partial r}\right)(\rho_0 w \operatorname{ctnh} \mu), \tag{3}$$

the tesseral component of pressure is associated with both horizontal and vertical motion. An exception would be the motion just at the surface of a perfectly spherical planet, where the vertical component would vanish, and, according to Eq. (19–2), the motion would be geostrophic. Even in this case, where there are no effects of terrain, the motion at an elevation above the spherical surface will not be exactly geostrophic. In view of the earlier estimates of the magnitude of w (Section 16) it can be foreseen that the second or thermobaric term in Eq. (19–2) will be quantitatively small in comparison to the first or geostrophic term. In another sense, however, the thermobaric motion is by no means negligible, since it determines the rise and subsidence of the air.

In order to estimate the magnitude of the longitudinal pressure gradient associated with the thermobaric motion, the planetary divergence equation, Eq. (19–3), may be simplified in two ways: first the term in $\partial/\partial\mu$ may be neglected in comparison with that in $r\dfrac{\partial}{\partial r}$; second, since $r = r_s + \chi$, $r\,\partial/\partial r$ may be replaced by $r_s\,\partial/\partial\chi$, etc. Then

$$\begin{aligned}\frac{\partial p_1}{\partial\lambda} &= \Omega r_s^{\,2}\tanh^2\mu[\partial(\rho_0 w)/\partial\chi] \\ &= \Omega r_s^{\,2}\sin^2\phi[\partial(\rho_0 w)/\partial\chi].\end{aligned} \tag{5}$$

To estimate orders of magnitude, suppose that $w = 1$ cm/sec, $\rho_0 = 10^{-3}$ g/cm³ at $\chi = 1$ km, while at $\chi = 0$, $w = 0$; the mean value of $\partial(\rho_0 w)/\partial\chi$ is then 10^{-8} c.g.s. units. Taking $\Omega r_s = 2$ km/sec and $\sin\phi = \frac{1}{2}$, the estimated pressure gradient is

$$\frac{1}{r_s\sin\phi}\frac{\partial p_1}{\partial\lambda} = 10^{-3}\ \text{dyne/cm}^3 = 0{\cdot}1\ \text{mb/km}.$$

This is a quite high value compared to those observed. It is therefore not unreasonable to assume that Eq. (19–4) can account for the observed gradients.

It is seen from Fig. 6 that the tesseral component of the Earth's pressure field consists of alternating high and low pressure areas arranged roughly along circles of latitude. Equation (19–4) shows that the gradient of vertical flow has the same sign as $\partial p_T/\partial\lambda$; and if $w = 0$ at $\chi = 0$, the same is true, in the lower atmosphere, of w itself. Hence the air rises over the western part of a high and the eastern part of a low; it sinks over the eastern part of a high and the western part of a low.

Moreover, since the atmosphere is stably stratified, w has the same sign as q, the net accession of heat. Hence the western part of a high and the eastern part of a low are regions of net heating, etc. These conclusions appear to be not inconsistent with the known facts.

The effects of terrain. In supposing that the surface of the planet is a perfectly smooth sphere, all effects of terrain have been ignored.

These are of two kinds: those caused by smooth variations in ground level and those caused by small obstacles, such as rocks and trees.

Consider, first, a smooth slope of 1 in 100, and suppose that a steady wind of 1 m/sec at ground level blows up this slope. This will require that the air have an upward velocity component of 1 cm/sec. It has been seen above that this is about the largest rate of rise that can be maintained by solar heating, and consequently, no *steady* wind in excess of 1 m/sec can be expected to blow up this slope. *Variable* winds of higher speeds can occur, since these can draw upon the heat capacity of the air or the latent heat of water vapour in the air.

It follows that even smoothly-variable terrain can have appreciable influence on the ground wind, and that this in turn must influence the ground-level values of the Earth's permanent pressure field. Such effects would appear as a fine structure of the isobars shown on Fig. 6. Apparently no attempt has been made to observe effects of this kind.

Smaller obstacles, which can be classed as roughness, will cause turbulence. The energy required for the production and maintenance of the turbulence must come largely from the large-scale motion of the air. This loss of energy to parasitic turbulence can be considered to be caused by a force retarding the large-scale motion, and hence vitiating the hypothesis of pure convection. This force doubtless has a very appreciable value in the lowest layer of the Earth's atmosphere. Its magnitude in the higher layers is unknown.

References for Chapter III

1. C. Abbé, Mechanics of the Earth's atmosphere, *Smithsonian Miscellaneous Collections* Vol. 51, No. 4, pp. 5, 7 (Washington, D.C., 1910).
2. V. Bjerknes, J. Bjerknes, H. Solberg and T. Bergeron, *Physikalische Hydrodynamik*, Springer, Berlin, 1933.
3. C. Eckart, Vortices and streams caused by sound waves, *Phys. Rev.* **73**, 68–76 (1948).
4. J. C. Freeman, Jr., L. Baer, and G. H. Jung, The bathystrophic storm tide, *J. Marine Research* **16**, 12 (1957).
5. G. Hadley, Concerning the cause of the general trade winds, *Phil. Trans. (London)* **39**, 58 (1735–36). (Reprinted as reference 1 above.)
6. H. Lamb, *Hydrodynamics*, Cambridge Univ. Press, 6th ed., 728 pp., 1932.
7. Rayleigh, *Scientific Papers*, No. 108, p. 246, Cambridge Univ. Press; *Phil. Trans. (London)* **175**, p. 1 (1883).
8. V. P. Starr, *Compendium of Meteorology*, p. 542, American Meteorological Society, Boston, 1951.

CHAPTER IV

THE FIELD EQUATIONS

20. Introduction

THE further development of the theory will be strongly dependent upon the use of suitable concepts and notations. This chapter will be largely devoted to the task of finding these tools.

The concept of energy is fundamental in all branches of science, and it will be well to begin by considering it in the present context. In Section 5, it was seen that the energy density of the fluid is

$$W = \tfrac{1}{2}\mathbf{u}^2 + \epsilon + g\chi \text{ erg/g}, \tag{1}$$

while the energy flow is

$$\mathbf{J} = p\mathbf{u} \text{ erg/cm}^2 \text{ sec}. \tag{2}$$

The equation of the conservation of energy was obtained as Eq. (5–6):

$$\rho \frac{DW}{Dt} + \nabla \cdot \mathbf{J} = \rho(\mathbf{f} \cdot \mathbf{u} + q). \tag{3}$$

When the various quantities entering into this equation are expanded into terms of different perturbation orders, it becomes

$$\rho_0 \frac{\partial W_0}{\partial t} = 0, \tag{4}$$

$$\rho_0 \frac{\partial W_1}{\partial t} + \rho_0 \mathbf{u}_1 \cdot \nabla W_0 + \nabla \cdot (p_0 \mathbf{u}_1) = \rho_0 q_1, \tag{5}$$

$$\rho_0 \frac{\partial W_2}{\partial t} + \rho_0(\mathbf{u}_1 \cdot \nabla W_1 + \mathbf{u}_2 \cdot \nabla W_0)$$
$$+ \rho_1 \frac{\partial W_1}{\partial t} + \rho_1 \mathbf{u}_1 \cdot \nabla W_0 + \nabla \cdot (p_1 \mathbf{u}_1 + p_0 \mathbf{u}_2)$$
$$= \rho_0(\mathbf{f}_1 \cdot \mathbf{u}_1 + q_2) + \rho_1 q_1. \tag{6}$$

Since W_0 depends only on the zero-order quantities, and they are independent of t, Eq. (20–4) is identically satisfied. Similarly W_1 depends only on the zero- and first-order quantities; explicitly, it is given by

$$W_1 = p_0 v_1 + \theta_0 \eta_1.$$

It is therefore easily verified that Eq. (20–5) is a direct consequence of the first-order equations of Section 9.

However, W_2 involves the second-order quantities ρ_2, $p_2 \ldots$, as well as squares and products of the first-order quantities ρ_1, p_1, $\mathbf{u}_1$ etc. Equation (20–6) cannot, therefore, be a consequence of the first-order equations, which do not involve $\rho_2, p_2 \ldots.$ This is an important conclusion, for one of the terms in W_2 is $\frac{1}{2}\mathbf{u}_1{}^2$, the kinetic energy (per gramme) of the first-order motion. One is therefore faced with three alternatives:

(1) To give up all thought of treating the first-order equations separately from those of higher orders;

(2) To give up, as a minimum, the use of the concept of kinetic energy in dealing with the first-order equations;

(3) To devise some way of using the kinetic energy concept that does not depend upon the use of Eq. (20–6), but only upon the first-order equations of Section 9.

The first two alternatives are unattractive; fortunately, it proves possible to adopt the third, as will be shown in the next section. In devising the required method, new concepts and notations will be suggested, that prove to be most useful in later work.

21. The external and thermobaric energies

If Eq. (9–2) is multiplied, term by term, with $\rho_0 \mathbf{u}_1$, Eq. (9–4) with $-\rho_0 p_0' Y_0 \eta_1/\eta_0' X_0$, Eq. (9–5) with $\rho_0 p_1/X_0$, and the results are added, many of the terms cancel. The resulting equation may be written

$$\frac{\partial E}{\partial t} + \nabla \cdot (p_1 \mathbf{u}_1) = \rho_0 \left\{ \mathbf{f}_1 \cdot \mathbf{u}_1 + p_1 q_1 \frac{Y_0}{\theta_0 X_0} - \frac{p_0' Y_0}{\eta_0' X_0} \frac{\eta_1 q_1}{\theta_0} \right\}, \qquad (1)$$

where

$$E = \tfrac{1}{2}\rho_0 \mathbf{u}_1{}^2 + \frac{p_1{}^2}{2\rho_0 c_0{}^2} - \tfrac{1}{2}\rho_0 \frac{p_0' Y_0}{\eta_0' X_0} \eta_1{}^2. \qquad (2)$$

Equation (21–1) has some points in common with Eq. (20–6): both contain the terms $\nabla \cdot (p_1 \mathbf{u}_1)$ and $\rho_0 \mathbf{f}_1 \cdot \mathbf{u}_1$, and E contains the kinetic energy term $\frac{1}{2}\rho_0 \mathbf{u}_1{}^2$. However, Eq. (21–1) and E contain no terms in

ρ_2, p_2, etc. and have been derived solely from the first-order equations.

Consequently, Eq. (21–1) is not the second-order energy equation, nor is E the same as $\rho_0 W_2$. It will appear that, for the present purposes, Eq. (21–1) is more useful than Eq. (20–6); and E, more useful than W_2. The reasons have been indicated in the Introduction and will appear more clearly below. In order to preserve distinctions when necessary, $\rho_0 W_2$ will be called the second-order energy density and E, the external energy density. The adjective "external" has only one advantage over others that might be suggested: it indicates that E has little relation to ε, the "internal" energy of thermodynamics. Where there is no danger of confusion, E will be called simply energy; similarly, $p_1\mathbf{u}_1$ is the external energy flow, but will be called simply energy flow, when no confusion is likely to be caused by omitting the adjective. The external energy is the sum of three terms; the first is the second-order kinetic energy, $\frac{1}{2}\rho_0\mathbf{u}_1^2$. The term $p_1^2/2\rho_0 c_0^2$ is familiar from acoustics,[1] where it is called the elastic energy. The term

$$-\tfrac{1}{2}\rho_0 \frac{p_0' Y_0}{\eta_0' X_0} \eta_1^2 = \tfrac{1}{2}\rho_0^2 g \frac{Y_0}{\eta_0' X_0} \eta_1^2$$

is unfamiliar. Since it involves both thermodynamic and gravitational quantities, it may be called the thermobaric energy.

It is seen that the thermobaric energy will have the same sign as the entropy gradient, η_0'. Since negative energies are usually associated with instability, this is another indication that the zero-order state will be stable or unstable according to the sign of the entropy gradient.

Using the formulae of Section 2, the thermobaric energy can be written in many different ways, two of which are

$$\tfrac{1}{2}\rho_0 g\left(\frac{\gamma_0 - 1}{a_0 c_0^2}\right)\frac{\eta_1^2}{\eta_0'} = \tfrac{1}{2}\rho_0 g\left(\frac{a_0\theta_0}{C_{p0}}\right)\frac{\eta_1^2}{\eta_0'}.$$

In the case of an ideal gas, these reduce to

$$\tfrac{1}{2}\rho_0 g \frac{\gamma - 1}{\gamma R}\frac{\eta_1^2}{\eta_0'},$$

and if its zero-order state is isothermal, to

$$\tfrac{1}{2}\rho_0\theta_0 \frac{\gamma - 1}{\gamma R}\eta_1^2.$$

Simpler formulae for the thermobaric energy will be derived below.

22. The field variables

In order to proceed further, it is essential to simplify the notation that has hitherto been used. Consider the external energy density

$$E = \tfrac{1}{2}\left(\rho_0 \mathbf{u}_1^{\,2} + \frac{p_1^{\,2}}{\rho_0 c_0^{\,2}} - \rho_0 \frac{p_0' Y_0}{\eta_0' X_0} \eta_1^{\,2}\right). \tag{1}$$

This may be written

$$E = \frac{1}{2c_0}(\mathbf{U}^2 + P^2 + N^2 Q^2) \tag{2}$$

if the definitions

$$\mathbf{U} = \mathbf{u}_1(\rho_0 c_0)^{1/2}, \quad P = p_1(\rho_0 c_0)^{-1/2}, \quad Q = (\eta_1/\eta_0')(\rho_0 c_0)^{1/2}, \tag{3}$$

$$N^2 = \rho_0 g \eta_0' Y_0 / X_0 = g\eta_0'(\gamma_0 - 1)/a_0 c_0^{\,2}, \tag{4}$$

are adopted. The quantities $\mathbf{U}$, P, Q will be called the field variables, and are to be used instead of $\mathbf{u}_1$, p_1, η_1 in most calculations. The coefficient N has the dimensions of a frequency and is a function only of χ. It will prove to be a most useful quantity in all calculations. The fields $\mathbf{U}$, P have the dimensions $\mathrm{g}^{1/2}\,\mathrm{sec}^{-3/2}$; Q has $\mathrm{g}^{1/2}\,\mathrm{sec}^{-1/2}$. The thermobaric energy is $N^2Q^2/2c_0$.

When the net accession of heat is zero, Eq. (9–12) shows that Q is proportional to the vertical displacement, h_1. Its interpretation as the first-order entropy, measured in suitable units, is more general. The other field variables are obviously measures of velocity and pressure, in variable units adapted to the nature of the problem.

In general, therefore, the units in which $\mathbf{U}$, P and Q measure the velocity, pressure and entropy will depend on the altitude χ, and are not constant. This disadvantage is more than offset by the simplification of the differential equations which results from their use. Examples of these variable units follow.

Oceanic case. The c.g.s. value of the units is determined by the acoustic impedance, $\rho_0 c_0$, and, in the case of Q, by the entropy gradient, η_0'. In the oceanic case, $(\rho_0 c_0)^{1/2}$ is very nearly constant, and varies from the value 380 c.g.s. only slightly; the corresponding value for fresh water is 375 c.g.s. The entropy gradient is highly variable and will be discussed in Sections 24 and 27.

Isothermal atmosphere. The variation of the acoustic impedance with altitude is given by the following equation where the subscript s indicates values at the surface $\chi = 0$:

$$(\rho_0 c_0)^{1/2} = (\rho_s c_s)^{1/2} \exp(-g\chi/2R\theta_0). \tag{5}$$

The entropy gradient is constant: $\eta_0' = g/\theta_0$.

Atmosphere with constant temperature gradient. In the notation of Section 8

$$(\rho_0 c_0)^{\frac{1}{2}} = (\rho_s c_s)^{\frac{1}{2}}[\chi_s/(\chi + \chi_s)]^{\frac{1}{2}\nu + \frac{1}{4}}, \tag{6}$$

while

$$\eta_0' = R[\nu + \gamma/(\gamma - 1)]/(\chi + \chi_s). \tag{7}$$

23. The field equations

When the first-order quantities are replaced by the fields, Eqs. (9–2, 4, 5) become

$$\frac{\partial \mathbf{U}}{\partial t} + c_0(\nabla + \Gamma\boldsymbol{\zeta})P - N^2 Q\boldsymbol{\zeta} + \boldsymbol{\Omega} \times \mathbf{U} = \mathbf{F}, \tag{1}$$

$$\frac{\partial P}{\partial t} + c_0(\nabla - \Gamma\boldsymbol{\zeta}) \cdot \mathbf{U} = G, \tag{2}$$

$$\frac{\partial Q}{\partial t} + \boldsymbol{\zeta} \cdot \mathbf{U} = H, \tag{3}$$

where

$$\Gamma = \frac{1}{2}\frac{(\rho_0 c_0)'}{\rho_0 c_0} + \frac{g}{c_0^{\,2}}, \tag{4}$$

$$\mathbf{F} = \mathbf{f}_1(\rho_0 c_0)^{1/2},$$
$$G = (Y_0 q_1/\theta_0)(\rho_0 c_0)^{-1/2}, \tag{5}$$
$$H = (q_1/\theta_0 \eta_0')(\rho_0 c_0)^{1/2}.$$

It will be noted that G and H are not independent:

$$G = (c_0 N^2/g)H. \tag{6}$$

The left-hand sides of the first three equations depend only on the three parameters N, Γ and c_0, and are otherwise much simpler than the original first-order equations. For example, the derivation of the external energy equation can be formulated as follows: multiply

Eqs. (23–1), (23–2) and (23–3) by **U**, P and N^2Q respectively; add the results and obtain

$$\frac{\partial E}{\partial t} + \nabla \cdot (P\mathbf{U}) = (\mathbf{F} \cdot \mathbf{U} + GP + N^2HQ)/c_0. \tag{7}$$

Incidentally, this establishes the important fact that the field equations are self-adjoint.

The boundary condition at a rigid surface remains essentially unaltered (cf. Eq. (11–1)):

$$\mathbf{n} \cdot \mathbf{U} = 0. \tag{8}$$

That at the free surface of an ocean or lake is (cf. Eq. (11–4))

$$c_0 \frac{\partial P}{\partial t} = g\mathbf{U} \cdot \boldsymbol{\zeta}, \tag{9}$$

and the displacement of its free surface is

$$h_1 = (Pc_0/g)(\rho_0 c_0)^{-1/2}. \tag{10}$$

The additive barotropic terms (cf. Section 10) are given by a less succinct equation than before. They are found by setting **U**, **F**, G, H and all time derivatives equal to zero. The field equations then reduce to

$$c_0(\nabla + \Gamma\boldsymbol{\zeta})P = N^2Q\boldsymbol{\zeta}, \tag{11}$$

which is satisfied when P is any differentiable function of χ, and

$$Q = (c_0/N^2)(P' + \Gamma P). \tag{12}$$

It is readily verified that this equation is equivalent to the less obscure equation

$$p_1' = -\rho_1 g$$

previously derived in Section 10.

In general, it will be found that the field equations are almost essential in dealing with time-varying solutions. To attempt to find these without the field notation leads to very cumbersome equations. On the other hand, steady-state solutions are usually more conveniently expressed in the notation of Chapters II and III.

24. Significance of the coefficients N and Γ

In order to obtain an insight into the significance of the parameters, it is necessary to find situations in which one or the other has the dominant dynamic effect. The velocity of sound is dominant in all acoustic phenomena; it will be supposed that the reader is sufficiently familiar with these so that no further discussion of c_0 is needed.

To obtain an interpretation of Γ, suppose that the level surfaces are plane, so that $\chi = z$, and $\boldsymbol{\zeta}$ is a constant, vertical, unit vector. Ignoring boundary conditions, let the fluid move with a steady vertical velocity: $\mathbf{u} = 0,0,w$, that is independent of time. Let $\mathbf{f}_1 = 0$, $g_1 = 0$; then the vertical motion will involve adiabatic changes of density, and w must be a function of z.

The conservation of matter requires that

$$\frac{\mathrm{d}}{\mathrm{d}z}(\rho w) = 0. \tag{1}$$

Since there is no heating, Eq. (5–3) becomes

$$\frac{\mathrm{d}\eta}{\mathrm{d}z} = 0. \tag{2}$$

The momentum equation, Eq. (5–1), reduces to

$$\frac{\mathrm{d}p}{\mathrm{d}z} = -\rho g. \tag{3}$$

Since $\mathrm{d}\eta/\mathrm{d}z = 0$, this may be written

$$c^2 \frac{\mathrm{d}\rho}{\mathrm{d}z} = -\rho g. \tag{4}$$

Equation (24–1) then yields

$$\frac{\mathrm{d}w}{\mathrm{d}z} = \frac{g}{c^2} w. \tag{5}$$

Introducing the field variable

$$W = w(\rho c)^{1/2},$$

Eq. (24–5) becomes

$$\frac{\mathrm{d}W}{\mathrm{d}z} - \Gamma W = 0. \tag{6}$$

This equation, involving Γ and W therefore represents the adiabatic expansion of the fluid under the specified ideal conditions.

A second equation involving Γ is most easily obtained from the energy equation, Eq. (10–6), which under these same conditions, reduces to

$$\frac{\mathrm{d}}{\mathrm{d}z}(PW) = 0. \tag{7}$$

Combining this with Eq. (24–6), the result is

$$\frac{\mathrm{d}P}{\mathrm{d}z} + \Gamma P = 0. \tag{8}$$

Equations (24–6) and (24–8) combine Γ with P, W, and their derivatives in a characteristic fashion that will become familiar in the following pages. Whenever these characteristic combinations appear, the reader will do well to think of adiabatic expansion.

The coefficient N is most readily understood by considering the oscillation of a small mass of the fluid when displaced from its zero-order position and then allowed to move freely. For definiteness, it is convenient to suppose this bit of fluid to be enclosed in a small balloon-like membrane—but this must be flaccid, so that the pressures inside and outside are always the same, and changes in density can occur unimpeded. It is again supposed that these occur adiabatically.

The balloon having been at the level z, is displaced to the level $z + \xi$, with the resulting change in its pressure

$$\delta p = -\rho g \xi. \tag{9}$$

The density inside the balloon has therefore changed by the amount

$$(\delta\rho)_{\text{inside}} = \delta p/c^2. \tag{10}$$

The density just outside the balloon was $\rho(z)$ before the displacement; after displacement, it is $\rho(z + \xi)$, or

$$(\delta\rho)_{\text{outside}} = \xi\, \mathrm{d}\rho/\mathrm{d}z. \tag{11}$$

The balloon therefore experiences a buoyant force (dynes/cm³) equal to

$$g[(\delta\rho)_{\text{outside}} - (\delta\rho)_{\text{inside}}]$$

$$= g\xi\left[\frac{\mathrm{d}\rho}{\mathrm{d}z} + \frac{\rho g}{c^2}\right]$$

$$= -\rho N^2\xi, \tag{12}$$

provided

$$N^2 = -g\left[\frac{1}{\rho}\frac{\mathrm{d}\rho}{\mathrm{d}z} + \frac{g}{c^2}\right]. \tag{13}$$

If Eq. (24–12) is equated to the inertial reaction of the balloon, its equation of motion is found to be

$$\frac{\mathrm{d}^2\xi}{dt^2} + N^2\xi = 0. \tag{14}$$

The balloon will therefore oscillate with the frequency N, given by Eq. (24–13), provided that frequency is real. If, however, N^2 is negative, the balloon will be unstable and depart widely from its initial position. This derivation of the frequency N is due to Väisälä;(3) it remains to be shown that Eq. (24–13) is identical with Eq. (14–4). This can be accomplished as follows: from Eq. (7–4),

$$\eta_0' = -\frac{a_0 c_0^2}{\gamma_0 - 1}\left(\frac{1}{\rho_0}\frac{\mathrm{d}\rho_0}{\mathrm{d}z} + \frac{g}{c_0^2}\right).$$

Substitution into Eq. (24–13) leads at once to

$$N^2 = \frac{(\gamma_0 - 1)g}{a_0 c_0^2}\,\eta_0', \tag{15}$$

which was to be shown.

One may therefore say that Γ summarizes the large-scale kinematics of the adiabatic expansion, while N summarizes its small-scale dynamics.

This formula for N^2 can be given other forms. According to Eq. (7–8), the adiabatic temperature gradient is

$$\theta_A' = \frac{g(\gamma_0 - 1)}{a_0 c_0^2} = \frac{g a_0 \theta_0}{C_{p0}},$$

so that

$$N^2 = \eta_0'\theta_A'. \tag{16}$$

Equation (7–9) expresses the entropy gradient in terms of the temperature gradient:

$$\eta_0' = (C_{p0}/\theta_0)(\theta_0' + \theta_A').$$

Hence

$$\begin{aligned} N^2 &= (C_{p0}\theta_A'/\theta_0)(\theta_0' + \theta_A'), \\ &= ga_0(\theta_0' + \theta_A'), \end{aligned} \tag{17}$$

a formula that is often most convenient.

25. Special formulae for the coefficients

Perfect gas. For dry air, which may be considered as a perfect gas

$$\begin{aligned} \gamma &= 1{\cdot}4, \\ a &= 1/\theta, \\ c^2 &= \gamma R\theta. \end{aligned}$$

Hence the adiabatic temperature gradient is

$$\theta_A' = g(\gamma - 1)/\gamma R, \tag{1}$$

and has the constant numerical value of 10°/km. The Väisälä frequency is therefore given by

$$N^2 = g(\theta_0' + \theta_A')/\theta_0. \tag{2}$$

The gas law can be rearranged into

$$\rho^2c^2 = \gamma p\rho = \gamma\rho^2/R\theta,$$

so that

$$\begin{aligned} \frac{(\rho c)'}{\rho c} &= \frac{p'}{p} - \frac{1}{2}\frac{\theta'}{\theta} \\ &= -\frac{\rho g}{p} - \frac{1}{2}\frac{\theta'}{\theta} \\ &= -\frac{g}{R\theta} - \frac{1}{2}\frac{\theta'}{\theta}, \end{aligned}$$

whence Eq. (23–4) yields

$$\Gamma = \left[\frac{2-\gamma}{\gamma-1}\theta_A' - \tfrac{1}{4}\theta_0'\right]\Big/\theta_0. \tag{3}$$

Fresh water. The approximations of Section 4 yield the expression

$$a = 15 \times 10^{-6} T_M$$

for the coefficient of thermal expansion of fresh water, where T_M is the temperature relative to the temperature of maximum density, Eq. (4–1). The specific heat may be taken as constant:

$$C_p = 1 \text{ cal/g}$$
$$= 4{\cdot}2 \times 10^7 \text{ erg/g},$$

whereupon the adiabatic temperature gradient is

$$\theta_A' = 3{\cdot}5 \times 10^{-10} T_M \theta_0 \text{ °C/cm}. \quad (4)$$

This yields $\theta_A' = 0{\cdot}11$°C/km at 15°C, $\rho = 1$ atm, and $\theta_A' = 0$ at the temperature of maximum density.

The Väisälä frequency is given by

$$N^2 = 1{\cdot}47 \times 10^{-2}(\theta_0' + \theta_A')T_M \quad (5)$$

and if $\rho c = 375$ is taken to be constant,

$$\Gamma = g/c^2 = 4{\cdot}4 \times 10^{-7} \text{ cm}^{-1}. \quad (6)$$

For all practical purposes, Γ may be set equal to zero, but the numerical value of N is not negligible.

Sea-water. The approximations of Section 4 yield the expression

$$a = 9{\cdot}3 \times 10^{-6} T_M$$

for the thermal expansion of sea-water of 35 parts per thousand. Hence the corresponding formulae for this fluid are

$$\theta_A' = 2{\cdot}2 \times 10^{-10} T_M \theta_0 \text{ °C/cm}, \quad (7)$$

$$N^2 = 0{\cdot}91 \times 10^{-2}(\theta_0' + \theta_A')T_M, \quad (8)$$

$$\Gamma = g/c^2 = 4{\cdot}4 \times 10^{-7} \text{ cm}^{-1}. \quad (9)$$

These values may be used for schematic discussions, but should not be used for the interpretation of empirical data. There are three reasons for this caution:

(1) The above formulae for thermal expansion do not adequately represent the accepted values, either at high or low pressures, but represent a compromise between precision and simplicity.

(2) The accepted values of thermal expansion are apparently quiet

inaccurate at low temperatures and high pressures, such as are encountered in the deep sea.

(3) It is physically obvious that salinity gradients (which have been ignored throughout the preceding) will affect the stability of the oceanic stratification. This is discussed by Hesselberg and Sverdrup,(2) who have published tables for the calculation of $E = N^2/g$ when salinity gradients are present.

In the next chapter, empirical values of N^2 for the ocean are presented. These have been determined with due regard for points (1) and (3). Unfortunately, it is impossible, at this time, to remedy the situation described under point. (2).

References for Chapter IV

1. P. M. Morse, *Vibration and Sound*, p. 237 McGraw-Hill, New York, 1948.
2. T. Hesselberg and H. V. Sverdrup, Die Stabilitäts-verhältnisse des Seewassers bei vertikalen Vershiebungen, *Bergens Museums Aarbok*, No. 15, 16 pp., 1915.
3. V. Väisälä, Uber die Wirkung der Windschwankungen auf die Pilot beobachtungen, *Soc. Sci. Fennica, Commentationes Phys.-Math.* II **19**, 37 (1925).

CHAPTER V

THE EARTH'S ATMOSPHERE, OCEANS AND LAKES

26. Introduction

THE resolution of an empirical state of fluid motion into zero- and higher-order terms is, to a very considerable extent, arbitrary. The perturbation calculation, if performed completely and correctly, should lead to the same end result no matter what zero-order state, within limits, is chosen as the starting point. However, the convergence of the successive approximations may be more rapid if one zero-order solution is adopted than if another is used; in fact, there may be some for which the calculations diverge—the existence of such states sets the limits mentioned above. In practice, only a few of the successive stages of the approximate calculations can be made. Unless the zero-order state is chosen so that the convergence is rapid, the degree of approximation achieved will then not be adequate. Unfortunately, there are few known principles upon which to base the choice.

In Chapter I, the commitment to a static, barotropic zero-order state has already been made. It therefore appears reasonable to look for such states that differ as little as possible from the observed states of the atmosphere or ocean, and to base the calculation on one of them. Since the observed states are rarely static and barotropic, one must resort to some form of averaging over space and time, such as was used in Chapter III. The nature of this average will depend upon the purpose of the proposed calculation.

If this purpose is climatological, one will average over the whole volume of the fluid concerned and over many seasons. For the discussion of more transitory phenomena, one will average over smaller volumes and shorter spans of time, hoping in this way to obtain more rapid convergence for the special phenomena under examination.

However, it is almost certain that such an average will not yield a static state: the averaged velocity will usually not be zero. One will then be confronted with the choice of abandoning the commitment to a static zero-order state, or else modifying the average in some way. Moreover, this modified average may not be barotropic, necessitating further modification in order to maintain the commitment. Consequently, it is scarcely possible to speak of an *empirical* zero-order state for any of the Earth's fluids.

These difficulties are to some extent theoretical; and if one adopts the principle that the zero-order state shall be that static, barotropic state which comes closest to some average state of the ocean or atmosphere, there is no practical difficulty in determining its general nature. The difficulties appear only when an attempt is made to specify it with great precision. Consequently, the following sections will discuss empirical zero-order states, but it is always to be understood that only the first significant figure is absolute. It is often necessary to use more significant figures in order to indicate differences in stability, etc. at various altitudes.

27. The stratification of the oceans

Large-scale averages. If seasonal changes are to be considered as first-order effects, the zero-order state must be determined by averaging over many seasons. There are some data of this sort for the oceans. In a general way, they confirm the barotropic distribution of the variables, but there are appreciable deviations. These must be eliminated by averaging, in some way, over latitude and longitude as well.

Figure 9 shows a graph of the variation of the velocity of sound with depth. It is necessarily somewhat schematic, but there are few places in the deep ocean that yield a markedly different graph. It will be noted that the scale is much exaggerated, the total variation being about three per cent. For many purposes, c_0 may be considered as constant. However, the minimum has marked effects on the long-range propagation of sound. This minimum is the result of the opposing effects of the temperature and pressure gradients; its depth is called the "axis" of the sound channel.

More striking are the variations in the coefficient N, shown by Fig. 10. The pronounced maximum is caused by the relatively high temperature gradient at a depth of several hundred meters. This region is known as the permanent thermocline. It will be noted that, at the thermocline, $2\pi/N$ has its minimum value of about twelve minutes. Everywhere above 2000 m, $2\pi/N$ is less than two hours. Below 2000 m, $2\pi/N$ increases in a roughly exponential fashion, if the existing data may be trusted, which is doubtful.

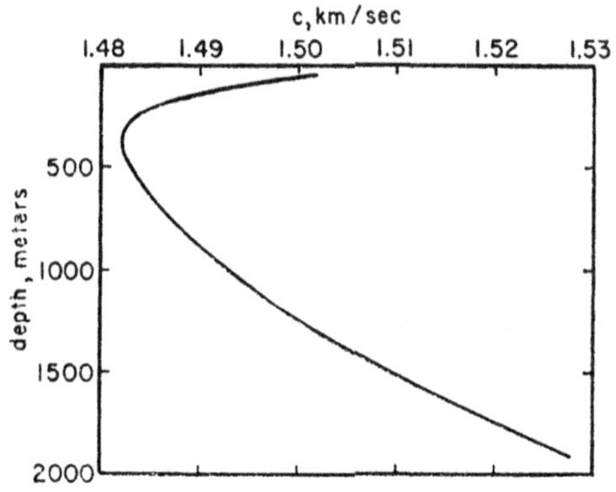

FIG. 9. The velocity of sound in the sea.

Temperature gradients at great depths are not measurable with great precision and, as has been noted, the value of the coefficient of thermal expansion is seriously in doubt for the range of pressures and

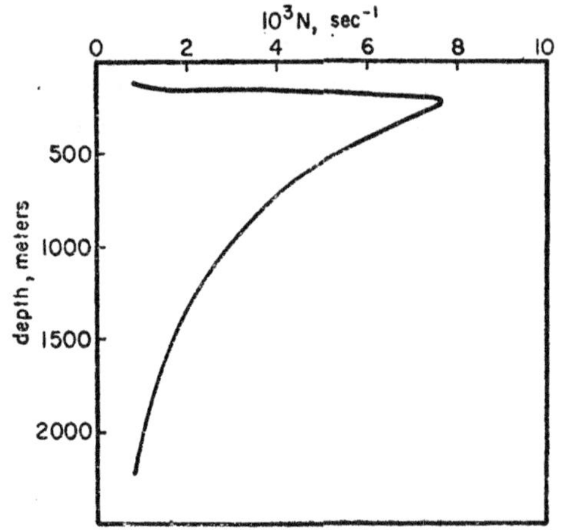

FIG. 10. The stability frequency, N, for the sea.

temperatures involved. The critical dimensionless parameter is the ratio N/Ω; the Coriolis force has a vertical component, and if $N/\Omega > 1$, the effects of this will be subordinated to those associated with the thermobaric forces. If $N/\Omega < 1$, the vertical Coriolis force will dominate over the thermobaric. These matters are treated more fully in Section 52.

TABLE 1

Values of N at 4000 m depth in the ocean for various temperature gradients and 0°C.

θ_0' (°C/km)	$10^8 N^2$ (sec^{-2})	$10^4 N$ (sec^{-1})	N/Ω
$-1 \cdot 0$	-139		
$-0 \cdot 1$	0	$0 \cdot 0$	$0 \cdot 0$
$0 \cdot 0$	$+14$	$3 \cdot 7$	$2 \cdot 6$
$+0 \cdot 1$	$+28$	$5 \cdot 2$	$3 \cdot 6$
$+1 \cdot 0$	$+141$	12	$8 \cdot 3$

Table 1 has been prepared with the aid of the approximate formulae of Section 25. It shows that if the temperature gradient is positive at great depths, $N/\Omega > 1$. However, a negative temperature gradient of even one degree in ten kilometers is sufficient to reduce the ratio to zero. This is the order of magnitude of the negative gradient observed below 5000 m depth. One may scarcely conclude that the ratio is greater than unity everywhere, and it would be most desirable to have better information on this point.

The coefficient Γ has such small values in the ocean that it may justifiably be set equal to zero in the equations. An estimate of its magnitude may be obtained from the formula

$$\Gamma = g/c^2 = 4 \cdot 5 \times 10^{-3} \text{ rad/km}$$

(Section 25) so that

$$2\pi/\Gamma = 1420 \text{ km/cycle.}$$

This magnitude is to be compared to the depth of the oceans, not to their horizontal extent. Since the mean depth is about $D = 5$ km, it follows that ΓD is only about 1/50 radian, and can thus be neglected for most purposes.

Small-scale averages. If the averages are extended over smaller regions of space and time, the foregoing description requires no change except at depths above 200 m. This upper layer is the scene of seasonal changes, and in the open ocean, often consists of an uppermost layer of almost isothermal water, with the stability $N^2 = ga_0\theta_A'$ (see Section 24). Assuming the high temperature of 30°C and a salinity of 35 parts per thousand, $N = 10^{-3}$ rad/sec. However, at 0°C, and the same salinity, $N = 1{\cdot}7 \times 10^{-4}$ rad/sec. Since $\Omega = 1{\cdot}45 \times 10^{-4}$ rad/sec, the ratio N/Ω will then depend critically on small temperature and salinity gradients. In general, the formulae of Section 25 are not sufficiently precise to calculate N from empirical data under these conditions.

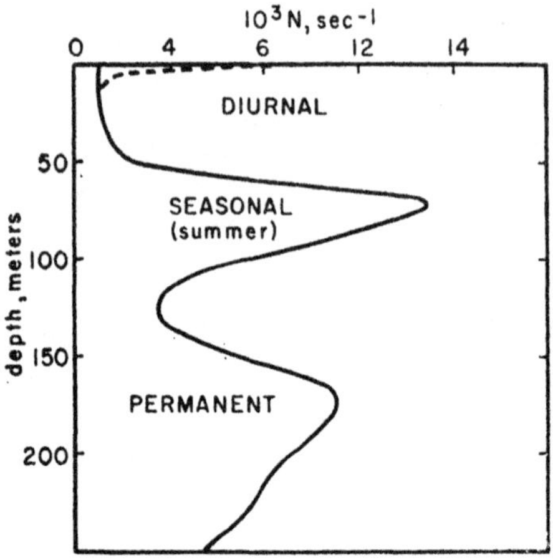

FIG. 11. Illustrating seasonal and diurnal changes in the stability of the upper layers of the sea.

This isothermal, or mixed layer, is most pronounced in winter and may then extend to the depth of the permanent thermocline. In summer, it is less extensive and may not exist at all. In the region between the mixed layer and the permanent thermocline, the temperature gradients are more pronounced, and may result in one or more maxima of N. These may be quite high, but are confined to layers of relatively small thickness, as illustrated in Fig. 11. Marked salinity gradients may also occur, which shift and distort the permanent thermocline.

With some idealization, the seasonal cycle may be described as follows: after the winter storms, the mixed layer is very deep, and beneath it, the stability rises abruptly to a maximum, decreasing monotonically beneath the thermocline. In late summer, after prolonged warm, relatively calm weather, the mixed layer is absent, and the stability increases irregularly with depth. It may show one or more pronounced maxima—the seasonal thermoclines—above the permanent thermocline. The stability at these shallow maxima may be very great, and is fully dissipated only after several storms and lower air temperatures have occurred in the fall.

In regions of low winds and high air temperatures, diurnal effects may also occur. These are usually highly stable surface layers that become pronounced during the early afternoon and disappear in the late evening.

28. The stratification of fresh-water lakes

The thermal state of fresh-water lakes is so strongly variable that zero-order states based on annual averages are largely useless. In general, averages extended over more than a few days are of doubtful value except in winter.

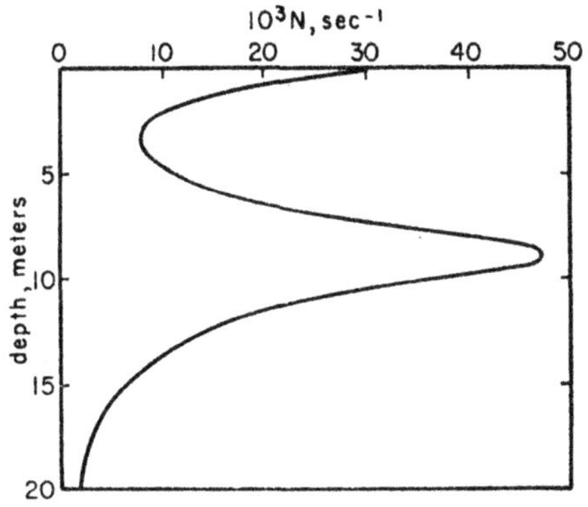

FIG. 12. Typical variation of the stability frequency with depth in a fresh-water lake (summer season).

During the summer months, lakes and reservoirs have a tendency to develop strong thermoclines, as illustrated on Fig. 12. A minimum of stability often, or usually, occurs between the thermocline and the surface. The typical values of N are five to ten times greater than in the ocean. There is considerable variation in this stratification; thus, the minimum stability above the thermocline may occur at the surface itself. The thermocline peak may be either lower and broader, or higher and narrower than that shown in the figure.

There appears to be a rough correlation between stability and size of the body of water, smaller bodies tending to be more stable. This extends even to tanks of water. In his classic work on turbulence, Reynolds took considerable care to use water that was as still and isothermal as possible. His stilling-tank was 18 in. deep and had a horizontal section of 18 by 72 in. Despite his care, the values of N ranged from 0·03 to 0·16 rad/sec in different experiments; it is certain that even larger values would have occurred in the absence of special precautions.

In the winter, this general correlation is overridden by another effect, which is associated with the temperature of maximum density of fresh water. On the first occasion that the air temperature drops below 4°C and is accompanied by wind, a complex transient phenomenon occurs. This is called overturn; its end result is to reduce the large vertical temperature gradients and to reduce the temperature of the whole volume of the lake to a value near that of maximum density. In the spring, the vertical temperature gradients are re-established somewhat more gradually.

Church(2) has studied these phenomena in Lake Michigan. In September 1942, he found the lake to be strongly stratified. There was a layer of small temperature gradient at the surface at about 20°C. The thermocline was at a depth of 10–20 m, and the temperature dropped to 10°C in a layer less than 5 m thick. Below this layer, the vertical gradient was still pronounced, with a temperature of 4·5°C at a depth of 70 m.

This contrasts strongly with his observations in February 1942: then the temperature at all points of the same section ranged only from 1·0 to 3·75°C. Moreover, the vertical temperature gradients were almost immeasurably small; and the coldest water was near shore, the warmest at the center of the lake. With little idealization, but also with scanty data, the isothermal surfaces may be described as cylinders erected on the bottom contours as bases. The inclination of the axes of these cylinders to the horizontal cannot be determined with great precision, but is almost certainly greater than 10°; this is suggestive of bathystrophic motion.

While the winter temperature of the lake is not exactly that of maximum density, it is clear that the smallness of both the vertical gradients and the coefficient of expansion will reduce the stability to a very small value. Table 2 has been based on the approximate formula of Section 25. An examination of this table and the empirical winter data on Lake Michigan suggests that N/Ω may be less than unity throughout a considerable part of its volume. One cannot avoid considering the possibility that the appropriate zero-order state may be that characterized by $T_m = 0$, $N = 0$, and that the observed departures from this state should be treated as first-, rather than zero-order effects.

TABLE 2

Values of N for freshwater lakes of various temperatures and temperature gradients

T_M (°C)	θ_0' (°C/km)	$10^8 N^2$ (sec^{-2})	$10^4 N$ (sec^{-1})	N/Ω
−2	−1	30	5·5	3·8
	0	0·6	0·8	0·5
	+1	−30	—	—
−1	−1	15	3·9	2·7
	0	0·15	0·4	0·3
	+1	−15	—	—
+1	−1	−15	—	—
	0	0·15	0·4	0·3
	+1	+15	3·9	2·7
+2	−1	−30	—	—
	0	0·6	0·8	0·5
	+1	+30	5·5	3·8

29. The stratification of the Earth's atmosphere

Figure 13a shows the variation of temperature with altitude in the Earth's atmosphere. The terminology for the various layers is not yet firmly established, but the region *Sm* corresponds roughly to the troposphere, *mM* to the stratosphere, *Mn* to the mesosphere, and $n\infty$ to the thermosphere. Different observers place the extrema at slightly differing altitudes, and also assign varying values to the extreme temperatures,(7) but all agree as to the general nature of the graph, except that some writers believe there are discontinuities in its slope.

Since $c^2 = \gamma R\theta$, the velocity of sound is indicated by the same graph. It is seen to vary within wider limits than in the ocean, and can scarcely be treated as a constant. Figure 13b shows the temperature gradient, as derived from Fig. 13a.

Figure 14 shows the variation of N^2 with altitude, calculated from the formulae of Section 25. It is seen that the maximum at A corresponds to a period of $2\pi/N = 4{\cdot}5$ min. This maximum stability in the stratosphere is thus rather greater than the stability of the ocean at the thermocline, but not by an order of magnitude. At the

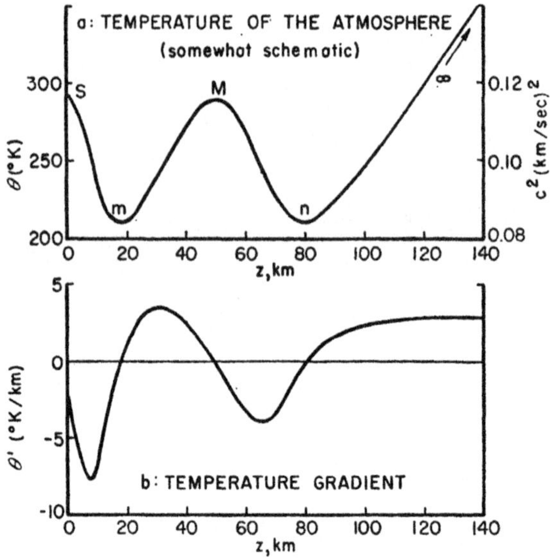

FIG. 13. Variation of temperature and temperature gradient with altitude in the Earth's atmosphere.

Earth's surface, the stability depends strongly on the method of averaging highly variable data, but is of the order of magnitude $N = 1$ rad/min.

Figure 14 also shows the variation of the parameter Γ; its values for the atmosphere are considerably larger than for the ocean; and since the atmosphere has an unlimited extent, it is not permissible to set Γ equal to zero without further examination of the consequences of such an assumption.

Although there are very few data for extremely high altitudes, and although it is certain that the air at such altitudes no longer obeys the laws of a continuous fluid, it is technically necessary to extrapolate the temperature to infinite altitudes. Figure 13 has been drawn to show a constant temperature gradient of 2·8°C/km above 120 km altitude. This is not inconsistent with present data, although the temperature increase cannot continue unaltered to infinity. This temperature gradient corresponds to an asymptotic pressure that varies inversely as the twelth power of the altitude (see Section 8).

Figure 15 shows the variation of pressure with altitude, as derived from the temperatures shown by Fig. 13. The question of the justification of treating rarefied gases as continuous fluids naturally arises in connection with the low pressures shown by this graph. At

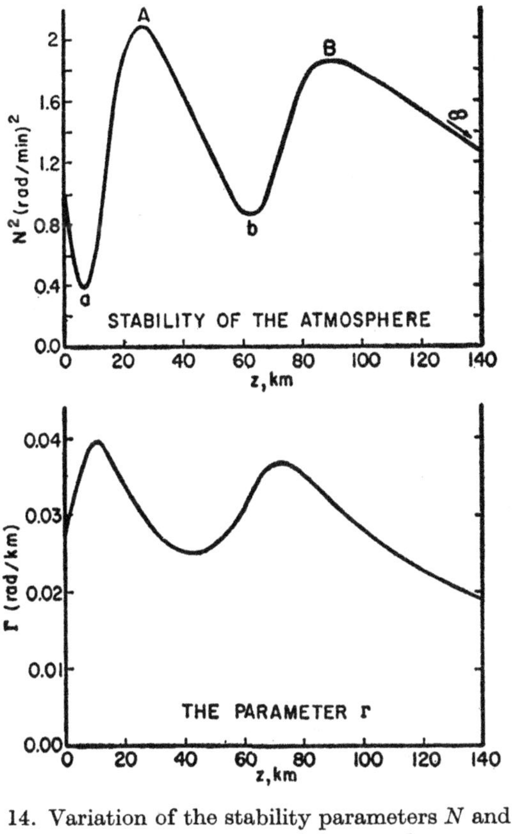

FIG. 14. Variation of the stability parameters N and Γ with altitude in the Earth's atmosphere.

an altitude of 100 km, the pressure is about 10^{-6} atm, and the mean free path of the air molecules is about 10 cm. It is therefore justifiable to use the continuum theory only if the scale of the phenomena is somewhat greater than 1 m. This leaves a considerable domain for the application of the theory even at this altitude. At greater altitudes, the minimum size will increase.

The kinematic viscosity of the air at 100 km is about 10^6 times greater than at sea level, and the same ratio applies to the other dissipative coefficients. These are very large values, but it is still justifiable to neglect them if the scale of the phenomenon is great enough.

At altitudes much greater than 200 km, it is doubtful whether the present theory is at all applicable. None the less, it is technically necessary to extend the solution of the equations to infinity, as has been remarked above. While this is admittedly an artificiality, it is preferable to "putting a lid on the atmosphere" and imposing even more artificial boundary conditions at a finite altitude.

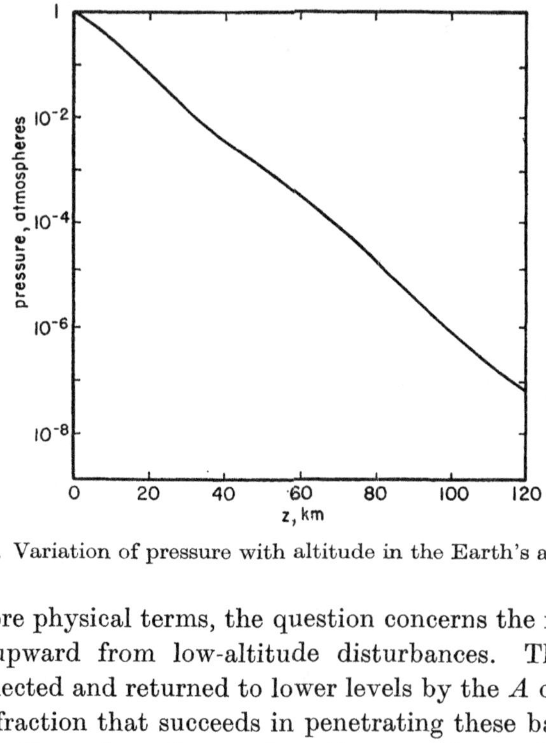

FIG. 15. Variation of pressure with altitude in the Earth's atmosphere.

In more physical terms, the question concerns the fate of energy radiated upward from low-altitude disturbances. This energy is largely reflected and returned to lower levels by the A or B maxima; the small fraction that succeeds in penetrating these barriers will be absorbed in the regions above, where the mean free path is very long. Absorption in these regions will affect phenomena at lower levels only slightly. Moreover, its effect will not be much different than if this energy were propagated to infinite altitudes without absorption. Such considerations will provide the clues to appropriate mathematical idealizations when the need for them arises.

30. Planetary rotation and cyclogenesis

Thus far, the rotation of the Earth has been introduced into the hydrodynamic equations without comment. The question arises whether the Coriolis force, $\mathbf{u} \times \mathbf{\Omega}$, exerts an appreciable influence on the motions. This term arises because, in the zero-order state, the fluids are at rest only in the co-ordinate system that rotates with the Earth.(3) In an inertial co-ordinate system, the fluid would be described as moving like a rigid body with the angular velocity $\mathbf{\Omega}/2$. In this section, the historical development of the ideas about the effect of the Earth's rotation on its atmosphere will be briefly sketched. In the next section, it will be shown that the field equations embody these ideas in a satisfactory mathematical form.

Early ideas. The importance of the Earth's rotation in meteorological and oceanographic phenomena has been repeatedly emphasized since very early times. Most of the earlier writers content themselves with establishing one simple fact: the relative velocity of two fixed points on the same meridian of the rotating Earth depends on their difference in latitude and can be very appreciable. The concept of Coriolis force was not introduced into the literature until after 1835(3) although Laplace(9) had derived the correct equations a few years earlier (*ca.* 1805); it is possible that they were known to Euler even before that.

Hadley,(5) in 1735, used this concept of relative motion to explain the trade winds. He supposed that, because of it, the streamlines of the ring vortices, discussed in Section 18, would be deflected in such a way as to give their sea-level velocity a westward component relative to the solid Earth. (A point on the equator of the solid Earth has an eastward velocity relative to other points on its meridian.) It very naturally escaped his attention that another effect of the Earth's rotation would be to unstabilize the vortices whose existence he postulated, since he did not investigate the dynamics of the problem.

Observations of Redfield and Reid: cyclones. At about the same time that the concept of the Coriolis force was being developed by physicists, an independent contribution to the subject was made by two acute observers. One of these, Redfield, is said to have observed the devastation caused by a hurricane that crossed New England. Making a journey through the area immediately following the storm, he noted that the uprooted trees gave evidence of the wind direction and correctly concluded that the air mass involved in the storm had a rotary motion. His observations and those of Reid were summarized by Charles Tracy[11] in 1843 as follows:

"The tendency of this evidence is to demonstrate that in the large storms which affect extensive districts, and also in the violent tornadoes which devastate a brief path, there are two motions, the rotary and progressive; and that the rotary is by far the most violent, and has a uniform direction of revolution, being from right to left if in the northern hemisphere, and the reverse if in the southern . . . . These propositions . . . have encountered doubts or gained a feeble faith in many minds for the want of a good cause to assign for the production of the alleged phenomena."

Tracy's theory. It is apparent that Tracy was unaware of the work of his predecessors for he advances an explanation that is essentially the same as Hadley's explanation of the trade-winds. The role of rising currents of air is emphasized: "the necessary condition, centripetal motion, may arise whenever a central spot subjected to intense heat, is surrounded by a cool atmosphere." The role of the Earth's rotation in deflecting the centripetal motion is clearly explained together with the comment: "in the northern states, a fire large enough to affect the atmosphere over a few acres may possess the essential force for generating a whirlwind, and may produce it in fact if the day be calm."

The emphasis upon upward and centripetal motions and the lack of reference to barometric disturbances may be called to the reader's attention. The importance of the latter was emphasized by Erman[4] (1868); however, Erman ignored the Earth's rotation.

Hann's theory. The ideas of the American workers did not immediately find acceptance in Europe. In 1875, they were championed by Julius Hann[6] who modified them considerably by including Erman's pressure gradients. He says: "the new theory, . . . finds in the formation of some local diminution of pressure the first cause of an inflow of air from all sides, and the attendant formation of a whirl in consequence of the Earth's daily rotation." While Hann's paper is largely non-mathematical it contains a formula analogous to the now commonplace "geostrophic formula" of Section 15. He considers a steady wind, of velocity v in the x-direction, and concludes that the isobaric surfaces will then slope in the y-direction.

His formula is essentially

$$g \frac{\partial h}{\partial y} = \Omega v \sin \phi,$$

h being the height of the surface and ϕ the latitude of the place under consideration. However, this formula is not used to explain the deflection of Tracy's centripetal currents but only "to make somewhat clearer the influence of the rotation of the Earth upon the origin of barometric differences."

Cyclones and anticyclones: vorticity. Tracy's rotary motions are now called cyclonic motions. The simplest illustration of a cyclone is the steady geostrophic motion of Section 15 when it occurs around a pressure minimum. If steady geostrophic motion occurs about a pressure maximum, it is called anticyclonic. However, cyclonic motion need not be steady. In modern times, the more general concept of vorticity is associated with the concept of a cyclone (or anticyclone).

Let u, v, w be the rectangular components of the velocity; then

$$\varpi = \frac{\partial v}{\partial x} - \frac{\partial u}{\partial y}$$

is the vertical component of the vorticity vector (cf. Section 14). In a cyclone, ϖ does not vanish; however, not all motions for which $\varpi \neq 0$ are cyclones or anticyclones. A non-zero value of ϖ is thus a necessary, but not a sufficient, condition for cyclonic or anticyclonic motion.

Helmholz and Bjerknes vorticity theorems. The concept of vorticity in connection with cyclonic motion derives its importance from two theorems concerning vorticity (Ref. 8, Chapter VII). It was first proven by Helmholz (1858) that if an incompressible non-viscous fluid is set into motion, starting from rest, by stirring or pouring, its motion will have zero vorticity. The exact relevance of this theorem to meteorological problems is not immediately apparent, but it does appear to indicate that cyclones could not easily be generated in such a fluid. Moreover, viscosity is not an important meteorological or oceanographic phenomenon. This suggests that cyclogenesis must be associated either with the compressibility of the geophysical fluids, or with their rotation.

The second theorem is usually attributed to V. Bjerknes (1918), and certainly it is because of his work that it has assumed a prominent place in meteorological literature. This theorem states, in essence, that whenever the state of a heavy compressible fluid is not barotropic (cf. Section 7), the time derivative of its vorticity will be different from zero.

The free motion of a heavy fluid is governed by the equation

$$\frac{\partial \mathbf{u}}{\partial t} + \mathbf{u} \cdot \nabla \mathbf{u} + \frac{1}{\rho} \nabla p + g \nabla \chi = 0, \tag{1}$$

if the Coriolis force be ignored. The term $\mathbf{u} \cdot \nabla \mathbf{u}$ is a complication in this equation and may also be ignored for the moment. Taking the curl of all terms, the result is then

$$\frac{\partial}{\partial t} (\nabla \times \mathbf{u}) = \frac{1}{\rho^2} \nabla \rho \times \nabla p. \tag{2}$$

The time rate of change of the vorticity, $\nabla \times \mathbf{u}$, is thus proportional to the vector product of the density and pressure gradients. For an incompressible fluid of constant density, $\nabla \rho = 0$, and so $\nabla \times \mathbf{u}$ is constant in time. In a barotropic state, the two gradients have the same (vertical) direction, and their vector product vanishes. When the two gradients have different directions, the state of the fluid is called baroclinic. In baroclinic states, the isobaric and isopyknic surfaces intersect.

When the terms $\mathbf{u} \cdot \nabla \mathbf{u}$ are taken into account, the above conclusion is complicated, but left essentially unchanged. (See also Section 14.) This baroclinic theory of cyclogenesis is strongly dependent on the non-linear terms

$$\nabla \times (\mathbf{u} \cdot \nabla \mathbf{u}) \qquad \text{and} \qquad \nabla \rho \times \nabla p,$$

and therefore presents great mathematical difficulties. It is certain that this type of cyclogenesis occurs in the atmosphere, especially in violent storms such as tornadoes. It is then complicated still further by precipitation.

However, it is clear that baroclinic cyclogenesis has little to do with the ideas of Hadley and Tracy. The essential point of their discussion is that the rotation of the Earth can cause cyclogenesis in another manner.

31. First-order cyclogenesis

The Tracy theory of cyclogenesis can most easily be discussed in terms of the field equations, and, for the first, the level surfaces will be supposed plane. Let U, V, W be the cartesian components of $\mathbf{U}$, W being vertical. In all cases, pure convection is assumed so that $\mathbf{F} = 0$.

No rotation. If $\Omega = 0$, the Eq. (23–1) is

$$\frac{\partial U}{\partial t} = -c\frac{\partial P}{\partial x}, \tag{1}$$

$$\frac{\partial V}{\partial t} = -c\frac{\partial P}{\partial y}, \tag{2}$$

$$\frac{\partial W}{\partial t} = -c\left(\frac{\partial P}{\partial z} + \Gamma P\right) + N^2 Q. \tag{3}$$

Redefining

$$\varpi = \frac{\partial V}{\partial x} - \frac{\partial U}{\partial y}, \tag{4}$$

the first two equations readily yield

$$\frac{\partial \varpi}{\partial t} = 0, \tag{5}$$

so that there is no first-order cyclogenesis in this case.

Rotation. If there is rotation, the axes may be chosen so that

$$\mathbf{\Omega} = \Omega_H \mathbf{j} + \Omega_V \mathbf{k} \tag{6}$$

and Eqs. (31–1) and (31–2) become

$$\begin{aligned} \frac{\partial U}{\partial t} - \Omega_V V + \Omega_H W &= -c\frac{\partial P}{\partial x}, \\ \frac{\partial V}{\partial t} + \Omega_V U &= -c\frac{\partial P}{\partial y}. \end{aligned} \tag{7}$$

On eliminating P by cross differentiation,

$$\frac{\partial \varpi}{\partial t} + \Omega_V\left(\frac{\partial U}{\partial x} + \frac{\partial V}{\partial y}\right) - \Omega_H \frac{\partial W}{\partial y} = 0. \tag{8}$$

In most cases, the term proportional to Ω_H will be small, since the vertical velocity is small, and hence the rate of increase of ϖ will be proportional to the horizontal convergence

$$-\left(\frac{\partial U}{\partial x} + \frac{\partial V}{\partial y}\right).$$

Just as the vorticity ϖ is a generalization of Tracy's rotary motion, so the horizontal convergence is a generalization of his centripetal motion.

Under these same conditions, the Eq. (23–2) is

$$\frac{\partial P}{\partial t} + c\left(\frac{\partial U}{\partial x} + \frac{\partial V}{\partial y} + \frac{\partial W}{\partial z} - \Gamma W\right) = G. \tag{9}$$

Again neglecting all terms in the vertical velocity W except $\partial W/\partial z$, Eqs. (31–8) and (31–9) yield

$$\frac{\partial \varpi}{\partial t} = \Omega_V\left(\frac{\partial P}{\partial t} + \frac{\partial W}{\partial z} - G\right). \tag{10}$$

Hence, heating, the vertical gradient of W, and pressure changes all generate vorticity in a rotating atmosphere.

Another relation between vorticity and horizontal convergence can be obtained from Eq. (31–7) by differentiation:

$$\frac{\partial}{\partial t}\left(\frac{\partial U}{\partial x} + \frac{\partial V}{\partial y}\right) - \Omega_V \varpi = -c\left(\frac{\partial^2 P}{\partial x^2} + \frac{\partial^2 P}{\partial y^2}\right). \tag{11}$$

Consequently both vorticity and horizontal convergence generate pressure gradients, in general accord with the ideas of Hann and Erman.

The horizontal convergence may be eliminated between Eqs. (31–8) and (31–11), yielding (with the usual neglect of W)

$$\frac{\partial^2 \varpi}{\partial t^2} + \Omega_V{}^2 \varpi = \Omega_V c\left(\frac{\partial^2 P}{\partial x^2} + \frac{\partial^2 P}{\partial y^2}\right), \tag{12}$$

which is a direct relation between cyclogenesis and horizontal barometric variations.

Rotation with spherical level surfaces. When the curvature of the level surfaces is taken into account, the above calculations are entirely analogous, but slightly more complicated. Using Mercator co-ordinates (cf. Appendix):

$$\begin{aligned} \Omega_H &= \Omega \operatorname{sech} \mu, \\ \Omega_V &= \Omega \tanh \mu. \end{aligned} \tag{13}$$

The appropriate definitions of the vertical vorticity ϖ and the horizontal convergence, Δ, are

$$\begin{aligned} \varpi &= \frac{\cosh \mu}{r}\left[\frac{\partial U_\nu}{\partial \lambda} - \cosh \mu \frac{\partial}{\partial \mu}(U_\epsilon \operatorname{sech} \mu)\right], \\ \Delta &= -\frac{\cosh \mu}{r}\left[\frac{\partial U_\epsilon}{\partial \lambda} + \cosh \mu \frac{\partial}{\partial \mu}(U_\nu \operatorname{sech} \mu)\right]. \end{aligned} \tag{14}$$

Equations (31–7) are replaced by

$$\begin{aligned} \frac{\partial U_\epsilon}{\partial t} - \Omega_V U_\nu + \Omega_H U_\zeta &= -\frac{c \cosh \mu}{r}\frac{\partial P}{\partial \lambda}, \\ \frac{\partial U_\nu}{\partial t} + \Omega_V U_\epsilon &= -\frac{c \cosh \mu}{r}\frac{\partial P}{\partial \mu}, \end{aligned} \tag{15}$$

and, after some algebra, the analogue of Eq. (31–8) is obtained as:

$$\frac{\partial \varpi}{\partial t} - \Omega_V \Delta + \frac{1}{r}\mathbf{\Omega} \cdot \mathbf{U} - \frac{\Omega_H}{r}\frac{\partial U_\zeta}{\partial \mu} = 0. \tag{16}$$

The additional term, $\mathbf{\Omega} \cdot \mathbf{U}/r$, is always very small compared to the other terms and may be ignored, together with the term in Ω_H.

Equation (31–9) is to be replaced by

$$\frac{\partial P}{\partial t} - c\Delta + c\left[\frac{1}{r^2}\frac{\partial}{\partial r}(r^2 U_\zeta) - \Gamma U_\zeta\right] = G, \tag{17}$$

whence the analogue of Eq. (31–10) is obtained in practically unaltered form.

The analogue of Eq. (31–11) is also obtained from Eqs. (31–15), in the form

$$\frac{\partial \Delta}{\partial t} + \Omega_V \varpi = -\frac{\Omega_H}{r}U_\epsilon - \frac{\Omega}{r}\frac{\partial U_\zeta}{\partial \lambda} + \frac{c \cosh^2 \mu}{r}\left[\frac{\partial^2 P}{\partial \lambda^2} + \frac{\partial^2 P}{\partial \mu^2}\right]. \tag{18}$$

Neglecting small terms, the analogue of Eq. (31–12) is then

$$\frac{\partial^2 \varpi}{\partial t} + \Omega_V{}^2 \varpi = \frac{c\,\Omega_V \cosh^2 \mu}{r}\left[\frac{\partial^2 P}{\partial \lambda^2} + \frac{\partial^2 P}{\partial \mu^2}\right]. \tag{19}$$

Summary. It has thus been shown that the classical ideas about cyclogenesis on a rotating Earth are implicit in the first-order field equations. Baroclinic cyclogenesis is not implicit in them, but would first appear in the secular equations obtained from a second-order calculation (cf. Section 14). In the case of mild, large-scale atmospheric phenomena, the baroclinic terms are likely to cause only small rates of change in ϖ, compared to those considered here. However, these secular changes may, over longer periods of time, be cumulative and distort the first-order motions considerably. As the intensity of the disturbance increases and its scale diminishes, the importance of these secular changes will doubtless increase.

References for Chapter V

1*a*. C. Abbé, The mechanics of the Earth's atmosphere, *Smithsonian Miscellaneous Collections*, Vol. 51, No. 4 (1910), Washington, D.C.

1*b*. C. Abbé, *Annual Report, Smithsonian Institution*, 1887, pp. 376–478, Washington, D.C.

2. P. E. Church, *The Annual Temperature Cycle of Lake Michigan*, Univ. of Chicago, Dept. of Meteorology: Part I, 1942; Part II, 1945.

3. G. Coriolis, Sur les équations du mouvement relatif des systèmes de corps, *J. École Polytechnique, Paris*, **5**, 142 (1835).

4. A. Erman, On the steady motion or the average condition of the Earth's atmosphere, *Astronomische Nachrichten*, **70**, 369 (1868). Translation in Ref. 1*a*, p. 31.

5. G. Hadley, Concerning the cause of the general trade winds, *Phil. Trans. (London)* **39**, 58 (1735). Reprinted in Ref. 1*a*, p. 5.

6. J. Hann, On the relation between the difference of pressure and the velocity of the wind . . . . , *Zeit. Austrian Meteorological Assoc.* **10**, 81 (1875). Translation in Ref. 1*b*, p. 426.

7. J. Kaplan, G. F. Schilling, and H. K. Kallman, Methods and results of upper atmosphere research, University of California, Institute of Geophysics; Scientific Report No. 3, 1954.

8. H. Lamb, *Hydrodynamics*, Cambridge Univ. Press, 6th ed., 728 pp., 1932.

9. P. S. Laplace, *Méchanique Céleste*, 4 vols. (Trans. by N. Bowditch), Boston, 1832. The relevant section is Part I, Book IV, Section 3, p. 543.

10. C. Tracy, On the rotary action of storms, *Am. J. Science and Arts* **45**, 65 (1843). Reprinted in Ref. 1*a*, p. 16.

CHAPTER VI

GENERAL THEOREMS CONCERNING THE FIELD EQUATIONS

32. Introduction

THIS chapter will be devoted to a brief summary of the most important theorems concerning the field equations of Section 23. These equations are generalizations of the wave equation

$$c^2 \nabla^2 \psi = \partial^2 \psi / \partial t^2.$$

It is assumed that the reader is familiar with the mathematical techniques for its solution, and will recognize the theorems as generalizations of those that apply to this equation.

The solutions of the field equations may be classified in several ways, and clarity demands some preliminary definitions. If the field variables **U**, P, Q are continuous and differentiable at all points occupied by the fluid, one speaks of a regular solution. Other solutions have somewhat the character of shock waves and are discontinuous on certain moving surfaces; strictly, these are irregular solutions, and it will simplify the discussion to adopt this strict definition of a regular solution. If the fluid extends to infinity, the further requirement that the field variables do not become infinite, even at great distances from the origin, will be imposed on regular solutions.

All regular solutions result in finite values of the external energy density and flux, as defined in Sections 21 and 22. If there is only a finite volume of the fluid, its total energy will also be finite; but if there is an infinite volume, its total energy may be infinite. Solutions corresponding to finite total energy will be called limited; the others, unlimited. This is a second classification. Unlimited solutions have certain paradoxes associated with them, which result from the possibility that energy is radiated to infinity and escapes from the finite parts of the fluid. It is assumed that the reader's familiarity with the wave equation will enable him to resolve these paradoxes when they arise.

33. The eigensolutions

The field equations are

$$\frac{\partial \mathbf{U}}{\partial t} + c(\nabla + \Gamma\boldsymbol{\zeta})P - N^2 Q\boldsymbol{\zeta} + \boldsymbol{\Omega} \times \mathbf{U} = \mathbf{F},$$
$$\frac{\partial P}{\partial t} + c(\nabla - \Gamma\boldsymbol{\zeta}) \cdot \mathbf{U} = G, \qquad (1)$$
$$\frac{\partial Q}{\partial t} + \boldsymbol{\zeta} \cdot \mathbf{U} = H,$$

$$\mathbf{n} \cdot \mathbf{U} = 0 \quad \text{on rigid boundaries,}$$
$$c(\partial P/\partial t) = g\boldsymbol{\zeta} \cdot \mathbf{U} \quad \text{on a free boundary.} \qquad (2)$$

The methods to be employed in the solution of this boundary-value problem depend on the fact that the left-hand sides of the above equations do not contain t explicitly in any of the coefficients c, N, Γ.

The free solutions, for which $\mathbf{F} = 0$, $G = 0$ and $H = 0$, therefore include some of the form

$$\mathbf{U} = \mathbf{U}_e e^{-i\omega t}, \quad P = P_e e^{-i\omega t}, \quad Q = Q_e e^{-i\omega t} \qquad (3)$$

where ω is a constant and $\mathbf{U}_e$, P_e, Q_e are independent of t. If these special solutions are regular (not necessarily limited), they are called eigensolutions. To avoid useless exceptions, the trivial solution $\mathbf{U}_e \equiv 0$, $P_e \equiv 0$, $Q_e \equiv 0$ will be excluded from further consideration and not included among the eigensolutions.

The eigensolutions therefore satisfy the equations

$$i\omega \mathbf{U}_e = c(\nabla + \Gamma\boldsymbol{\zeta})P_e - N^2 Q_e \boldsymbol{\zeta} + \boldsymbol{\Omega} \times \mathbf{U}_e,$$
$$i\omega P_e = c(\nabla - \Gamma\boldsymbol{\zeta}) \cdot \mathbf{U}_e, \qquad (4)$$
$$i\omega Q_e = \boldsymbol{\zeta} \cdot \mathbf{U}_e,$$

$$\mathbf{n} \cdot \mathbf{U}_e = 0 \quad \text{on rigid boundaries,}$$
$$g\boldsymbol{\zeta} \cdot \mathbf{U}_e + i\omega c P_e = 0 \quad \text{on a free boundary.} \qquad (5)$$

It is to be noted that the boundary conditions are included in the definition of an eigensolution, as is the requirement of regularity.

In general, these last five equations will have no regular solution (other than the trivial one); they will have a non-trivial solution only for certain values of ω. These are the eigenfrequencies or eigenvalues. If the parameter $N^2 > 0$ at all points occupied by the fluid, the eigenvalues will all be real; if $N^2 < 0$ at some points in the fluid, some of the eigenvalues may be imaginary or complex. In what follows, it will always be supposed that $N^2 > 0$ at all points of the fluid, so that only real eigenvalues need be considered.

The eigensolutions will generally be complex functions of position, since i appears explicitly in Eqs. (33–4) and (33–5). It is easily seen that if $\mathbf{U}_e$, P_e, Q_e is an eigensolution associated to ω, then $\mathbf{U}_e^*$, P_e^*, Q_e^* (the asterisk indicates the complex conjugate) will be a solution for $-\omega$. Hence if ω is an eigenvalue, $-\omega$ will also be one. This sometimes enables one to ignore negative values of ω. Since the eigensolutions are complex, it becomes necessary to modify the definitions of energy density and energy flux; the most useful definitions are

$$E = \frac{1}{2c}\left[|\mathbf{U}_e|^2 + |P_e|^2 + N^2|Q_e|^2\right], \tag{6}$$

$$J = \tfrac{1}{2}[P_e\mathbf{U}_e^* + P_e^*\mathbf{U}_e]. \tag{7}$$

Let a be any numerical constant, and the remaining notation the same as in the last paragraph. Then $a\mathbf{U}_e$, aP_e, aQ_e will be a second eigensolution associated to ω. It is convenient not to consider these as different; more precisely, one will select one as representative of all and ignore the rest. If the eigensolution is limited, this representative will be chosen so that the total energy of the fluid is unity. In the case of unlimited eigensolutions, some other method of choosing the representative must be adopted. In all that follows, this representative will be called *the* eigensolution associated to ω.

34. The expansion theorem

The eigensolutions derive their importance from the fact that the most general limited solution of Eqs. (33–1, 2) can be constructed out of them. Moreover, only summations and integrations are needed for this construction.

The theorem takes on different forms, the simplest applying to the case of a finite volume of fluid. In this case, it is analogous to the Fourier series expansion of a function. The eigensolutions are then denumerable—i.e., they may be numbered with an index, n, that takes

7

on only integral values. Let $\mathbf{U}_n$, P_n, Q_n be the eigensolution associated to the eigenfrequency ω_n; a preliminary theorem states that the eigensolutions may be chosen so that

$$\iiint \{[\mathbf{U}_n{}^*\mathbf{U}_m + P_n{}^*P_m + N^2Q_n{}^*Q_m]/2c\}\, dx\, dy\, dz$$

$$= 0 \quad \text{if} \quad n \neq m, \tag{1}$$

$$= 1 \quad \text{if} \quad n = m.$$

In this equation, the integral is to be extended over the whole volume occupied by the fluid; it is called the orthogonality theorem. In terms of these eigensolutions, one may state the

Expansion theorem (for a finite volume). Every regular solution of Eqs. (33–1, 2) has the form

$$\mathbf{U} = \Sigma_n A_n(t)\mathbf{U}_n(x, y, z),$$

$$P = \Sigma_n A_n(t) P_n(x, y, z), \tag{2}$$

$$Q = \Sigma_n A_n(t) Q_n(x, y, z);$$

moreover,

$$(dA_n/dt) + i\omega_n A_n = B_n(t), \tag{3}$$

where

$$B_n = \iiint \{[\mathbf{U}_n{}^* \cdot \mathbf{F} + P_n{}^*\ G + N^2Q_n{}^*H]/2c\}\, dx\, dy\, dz. \tag{4}$$

It is to be noted that the same function $A_n(t)$ appears in all three of Eqs. (34–2), and that it is related in a definite way to the generalized forces $\mathbf{F}$, G, H by Eqs. (34–3) and (34–4). If these forces vanish (free motion), the solution of Eq. (34–3) is

$$A_n(t) = C_n e^{-i\omega_n t},$$

C_n being a constant of integration. The general solution of Eq. (34–3) is only slightly more elaborate.

It may also be remarked that, while the expansion theorem is asserted only for regular solutions, it also applies to *some* irregular solutions that represent shock waves, storm fronts, point sources of energy, etc. It does not apply to all irregular solutions.

When the volume of the fluid is infinite, these theorems become much more elaborate, for reasons that are connected with the radiation of energy. A finite volume of fluid is necessarily bounded by a closed surface, and the equations imply that energy is totally reflected at this boundary. When the volume of fluid is infinite, its boundary is necessarily open, and it *may* be possible for energy to be radiated to infinity. It will be seen below that this is *not* possible for certain zero-order states, even though the fluid extends to infinity; in these cases, the rays are refracted so that energy is always returned to the level from which its starts. For these zero-order states, all eigensolutions are limited, and the above theorems apply without essential modification.

However, for many zero-order states (including the isothermal atmosphere of ideal gas) energy can be radiated to infinite altitudes, unless a physical boundary reflects it. When the fluid is physically unbounded, some or all of the eigensolutions are unlimited, and the theorems become analogous to the Fourier integral expansion, or to a combination of a series and an integral.

In the simplest of these cases, the eigensolutions are all unlimited, and they may be identified by a single, continuously variable parameter, ν, as may the eigenfrequencies. Let

$$\mathbf{U}(x, y, z, \nu) \equiv \mathbf{U}(\nu), \quad \text{etc.},$$

be associated to the eigenfrequency $\omega(\nu)$. Then the orthogonality theorem states that these functions may be chosen so that

$$\iiint \{[\mathbf{U}^*(\nu) \cdot \mathbf{U}(\mu) + P^*(\nu)P(\mu) + N^2Q^*(\nu)Q(\mu)]/2c\}\, dx\, dy\, dz$$
$$= \delta(\nu - \mu), \tag{5}$$

where $\delta(x)$ is Dirac's delta function; in particular, $\delta(0) = \infty$, so that the unlimited nature of the eigensolutions is made explicit by this equation. In terms of these functions, one obtains the

Expansion theorem (for an infinite volume). Every regular and limited solution of Eqs. (33–1, 2) has the form

$$\mathbf{U} = \int A(t, \nu)\mathbf{U}(x, y, z, \nu)\, d\nu,$$

$$P = \int A(t, \nu)P(x, y, z, \nu)\, d\nu, \tag{6}$$

$$Q = \int A(t, \nu)Q(x, y, z, \nu)\, d\nu;$$

moreover

$$(dA(t, \nu)/dt) + i\omega(\nu)A(\nu, t) = B(\nu, t), \tag{7}$$

where

$$B(\nu, t) = \iiint \{[\mathbf{U}^*(\nu) \cdot \mathbf{F} + P^*(\nu)G + N^2Q^*(\nu)H]/2c\}\, dx\, dy\, dz. \tag{8}$$

The analogy to the case of a finite volume will be apparent without detailed comment. It is to be noted, however, that the *unlimited* eigensolutions are needed to construct the *limited* solutions of Eqs. (33–1, 2). This makes it essential to consider unlimited eigensolutions, if and when they arise. This conclusion is the principal reason for including this chapter here; there has been considerable and inconclusive discussion of the justification for considering the unlimited eigensolutions in meteorology.

Equations (34–5) to (34–8) apply only to the special case in which all eigensolutions are unlimited, and all can be identified with a single parameter. More complicated cases can arise; some eigensolutions may be limited, some unlimited; it may be necessary to use more than one parameter to identify the unlimited ones, or even parameters and integers combined. These complexities need not be examined in detail, however.

CHAPTER VII

FORMULATION OF THE MAJOR MATHEMATICAL PROBLEMS

35. Introduction

WITH the derivation of the field equations, the general theory of first-order effects has been implicitly formulated. It remains to examine it in systematic detail and to apply it to special cases. Essentially, this would involve the solution of the field equations for a representative variety of boundary conditions, driving forces, and zero-order states. At the present time, it would be impossible to exhaust all the possibilities that immediately present themselves, and it is necessary to be content with a very limited selection from them.

The first decision here will be to select the eigensolutions for further study and to omit all other problems. This is made reasonable by the discussion in Chapter VI; however, the results of Chapter III indicate that many interesting and important problems are thereby excluded. Henceforth, the eigensolutions will be indicated simply by **U**, P, Q, with the omission of the subscripts used in the last chapter. These quantities will then satisfy the equations

$$i\omega \mathbf{U} - \boldsymbol{\Omega} \times \mathbf{U} = c(\nabla + \Gamma\boldsymbol{\zeta})P - N^2 Q\boldsymbol{\zeta}, \tag{1}$$

$$i\omega P = c(\nabla - \Gamma\boldsymbol{\zeta}) \cdot \mathbf{U}, \tag{2}$$

$$i\omega Q = \boldsymbol{\zeta} \cdot \mathbf{U}. \tag{3}$$

The second decision will be to consider only those cases in which the fluid is bounded only by level surfaces. This also excludes very important problems, such as the effect of the continents on oceanic currents and tides, the effect of terrain on atmospheric phenomena, etc. Some attempt will be made to mitigate the effects of this decision by sections devoted to especially simple problems, but in this chapter it will not be reconsidered. Because of this decision, the boundary conditions need not always be made explicit.

For the limited class of problems thus defined, the solution of the three equations quoted can be reduced to the solution of simpler equations that will be derived in this chapter; their more or less detailed solution will then be considered in later chapters.

It will first be shown that if the fields P and Q can be determined, the velocity $\mathbf{U}$ can be calculated from formulae that involve only differentiation and algebraic operations. These two fields take the place of the potentials used in classical hydrodynamics.

It will then be shown that these two fields can be found from the solution of simpler equations. The first set of these is a pair of ordinary differential equations, with altitude as the independent variable. They are conveniently called residual equations and always have the form of Eq. (36–12).

The second differential equation is partial, involving the horizontal co-ordinates. It takes on different forms under different circumstances. If there is no rotation and the level surfaces are treated as planes, it is the usual two-dimensional wave equation, Eq. (36–11). If there is no rotation and the level surfaces are spheres, it is Eq. (36–23), and may be called the wave equation for a spherical surface. In Chapter XVI it will be shown that its solution can be expressed in terms of Legendre polynomials.

When the Coriolis force cannot be neglected, these problems become much more difficult. They were first encountered by Laplace in his theory of tides.(3) Concerning them, Hough(2) remarks, "the theory . . . originally presented by Laplace in the *Mécanique Céleste* . . . has been described by Airy as 'perhaps more obscure than any other part of the same extent in that work.' The obscurity complained of does not, however, seem to have been entirely removed by Laplace's successors."

These difficulties and obscurities have continued to discourage the development of this fundamental part of oceanographic and meteorological theory, and the lack of this formulation has undoubtedly retarded the development of these sciences during the past century. The difficulties are real and cannot all be overcome at this time. The obscurities that encumbered Laplace's pioneer work are no longer excusable, even though the subject does not lend itself to a brief exposition.

36. The case of no rotation

Plane level surfaces. If $\mathbf{\Omega} = 0$, and $\boldsymbol{\zeta} = \mathbf{k}$, a constant, cartesian co-ordinates may be introduced with the z-axis along $\boldsymbol{\zeta}$. If U, V, W are the components of $\mathbf{U}$, Eqs. (35–1, 2, 3) become

$$i\omega U = c\frac{\partial P}{\partial x}, \tag{1}$$

$$i\omega V = c\frac{\partial P}{\partial y}, \tag{2}$$

$$i\omega W = c\left(\frac{\partial P}{\partial z} + \Gamma P\right) - N^2 Q, \tag{3}$$

$$i\omega P = c\left(\frac{\partial U}{\partial x} + \frac{\partial V}{\partial y} + \frac{\partial W}{\partial z} - \Gamma W\right), \tag{4}$$

$$i\omega Q = W. \tag{5}$$

The first two equations show that the horizontal components of the velocity are proportional to the horizontal pressure gradient, and the fifth shows that the amplitude of the vertical displacement is $|Q|$, when measured in the present system of units. These statements may by summarized by the equation

$$\mathbf{U} = (c/i\omega)\nabla_H P + i\omega Q\mathbf{k}, \tag{6}$$

if the notation

$$\nabla_H = \mathbf{i}\frac{\partial}{\partial x} + \mathbf{j}\frac{\partial}{\partial y} \tag{7}$$

is adopted. The first term of the sixth equation may be called the gradient velocity; the second, the thermobaric.

Residual equations. The velocity is thus determined once the P and Q fields have been determined. However, in the absence of detailed knowledge of these, the relative magnitudes of the horizontal gradient velocity and the vertical thermobaric velocity cannot be estimated. If the velocity component W is eliminated between Eqs. (36–3) and (36–5), the equation

$$c\left(\frac{\partial P}{\partial z} + \Gamma P\right) = (N^2 - \omega^2)Q \tag{8}$$

is obtained. This involves only P and Q, and will be called the first residual equation. A second residual equation is obtained on eliminating the velocity components from Eq. (36–4) by substitution from the other equations:

$$c\left(\frac{\partial Q}{\partial z} - PQ\right) = P + \frac{c^2}{\omega^2}\left(\frac{\partial^2 P}{\partial x^2} + \frac{\partial^2 P}{\partial y^2}\right). \tag{9}$$

Separation of variables: the two-dimensional wave equation. In the residual equations, the independent variables are well separated, since none of their coefficients depend on x or y. Consequently, they will have special solutions of the form

$$P = P_2(z)P_1(x, y), \qquad Q = Q_2(z)Q_1(x, y). \tag{10}$$

In order that this equation represent a solution of Eqs. (36–8) and (36–9), the factors must satisfy the equations

$$\frac{\partial^2 P_1}{\partial x^2} + \frac{\partial^2 P_1}{\partial y^2} + \varkappa^2 P_1 = 0, \quad P_1 = Q_1 \tag{11}$$

$$\begin{aligned} c\left(\frac{\partial P_2}{\partial z} + \Gamma P_2\right) &= (N^2 - \omega^2)Q_2, \\ c\left(\frac{\partial Q_2}{\partial z} - \Gamma Q_2\right) &= (1 - \varkappa^2 c^2/\omega^2)P_2, \end{aligned} \tag{12}$$

where $\varkappa^2$ is a constant, called the constant of separation.

Equations (36–12) are ordinary differential equations; their theory will form the subject matter of later chapters. Equation (36–11) is known as the two-dimensional wave equation; only its simplest solutions will be needed for present purposes; these will be discussed in Chapter VIII.

Spherical level surfaces. In this case, the Eqs. (35–1, 2, 3) become

$$i\omega U_\epsilon = (c/r)\cosh\mu\,\frac{\partial P}{\partial\lambda}, \tag{13}$$

$$i\omega U_\nu = (c/r)\cosh\mu\,\frac{\partial P}{\partial\mu}, \tag{14}$$

$$i\omega U_\zeta = c\left(\frac{\partial P}{\partial r} + \Gamma P\right) - N^2Q, \tag{15}$$

$$i\omega P = (c/r)\cosh\mu\left[\frac{\partial U_\epsilon}{\partial\lambda} + \cosh\mu\,\frac{\partial}{\partial\mu}(U_\nu\operatorname{sech}\mu)\right] + c\left[\frac{1}{r^2}\frac{\partial}{\partial r}(r^2U_\zeta) - \Gamma U_\zeta\right], \tag{16}$$

$$i\omega Q = U_\zeta, \tag{17}$$

provided Mercator co-ordinates are used. Inspection of Eqs. (36–13), (36–14) and (36–17) shows that Eq. (36–6) remains valid, with the notation

$$\nabla_H = (1/r)\cosh\mu\left[\boldsymbol{\epsilon}\,\frac{\partial}{\partial\lambda} + \boldsymbol{\nu}\,\frac{\partial}{\partial\mu}\right].$$

The velocity again consists of the horizontal gradient component and the vertical thermobaric.

Residual equations. Substitution of Eq. (36–17) into Eq. (36–15) yields the first residual equation as

$$c\left(\frac{\partial P}{\partial r} + \Gamma P\right) = (N^2 - \omega^2)Q, \tag{18}$$

and substitution into Eq. (36–16) yields the second as

$$c\left[\frac{1}{r^2}\frac{\partial}{\partial r}(r^2Q) - \Gamma Q\right] = P + \frac{c^2\cosh^2\mu}{\omega^2r^2}\left(\frac{\partial^2P}{\partial\lambda^2} + \frac{\partial^2P}{\partial\mu^2}\right). \tag{19}$$

Equation (36–18) is practically identical with Eq. (36–8), and the resemblance of Eqs. (36–9) and (36–19) will be obvious.

The resemblance can be further increased by an approximation that is sufficient for most purposes: since

$$r = r_s + \chi$$

where r_s is the radius of mean sea level and χ the altitude above mean sea level; and moreover since $r_s \gg \chi$ in all regions of importance, the Eqs. (36–18) and (36–19) may be written, approximately,

$$c\left(\frac{\partial P}{\partial \chi} + \Gamma P\right) = (N^2 - \omega^2)Q, \tag{20}$$

$$c\left(\frac{\partial Q}{\partial \chi} - \Gamma Q\right) = P + \frac{c^2 \cosh^2 \mu}{\omega^2 r_s^{\,2}}\left(\frac{\partial^2 P}{\partial \lambda^2} + \frac{\partial^2 P}{\partial \mu^2}\right). \tag{21}$$

Separation of variables: wave equation for a spherical surface. In the last two equations, the independent variables are again well separated; and they will have special solutions of the form

$$P = P_1(\lambda, \mu)P_2(\chi), Q = Q_1(\lambda, \mu)Q_2(\chi). \tag{22}$$

The functions P_2 and Q_2 must again satisfy Eq. (36–12), with χ replacing z. Consequently, these equations are applicable regardless of the nature of the level surfaces.

With spherical level surfaces, P_1 and Q_1 must satisfy the equation

$$\cosh^2 \mu\left(\frac{\partial^2 P_1}{\partial \lambda^2} + \frac{\partial^2 P_1}{\partial \mu^2}\right) + \varkappa^2 r_s^{\,2} P_1 = 0, \quad P_1 = Q_1, \tag{23}$$

which is the wave equation for a spherical surface. The solution of this equation was worked out by Legendre and will be discussed in Chapter XVI.

37. Rotation with plane level surfaces

Three phenomena are basic factors in the hydrodynamics of the geophysical fluids:

(1) Their stratification,
(2) the curvature of the level surfaces,
(3) the rotation of the Earth (or planet).

It requires little observational material to establish the existence of these three factors, but their dynamic interaction in the fluid is so complex that even large amounts of observational material are of little help in understanding the situation. A theoretical analysis appears to be the only way to clarify it.

The previous section has given a preliminary account of the mathematical theory when the rotation is ignored. This and the next section will provide a similar survey of the problems that arise when the Coriolis force is included in the equations. The present section ignores the complication of curved level surfaces, which may be justifiable for small-scale phenomena, and in any event, provides a useful introduction to the more realistic case of spherical level surfaces.

Using the same notation as in Section 36 and choosing the y-axis toward the north so that

$$\mathbf{\Omega} = \Omega_H \mathbf{j} + \Omega_V \mathbf{k}, \tag{1}$$

the equations to be solved are (cf. Eqs. (36–1, 2, 3))

$$i\omega U + \Omega_V V - \Omega_H W = c\,\frac{\partial P}{\partial x}, \tag{2}$$

$$i\omega V - \Omega_V U = c\,\frac{\partial P}{\partial y}, \tag{3}$$

$$i\omega W + \Omega_H U = c\left(\frac{\partial P}{\partial z} + \Gamma P\right) - N^2 Q, \tag{4}$$

$$i\omega P = c\left[\frac{\partial U}{\partial x} + \frac{\partial V}{\partial y} + \left(\frac{\partial W}{\partial z} - \Gamma W\right)\right], \tag{5}$$

$$i\omega Q = W. \tag{6}$$

The traditional approximations. All published treatments of these equations involve approximations. Because of the obscurities already mentioned, it is not always possible to identify these approximations with certainty. This obscurity results from two circumstances: the writers have not always been careful to derive the equations correctly, and they have not used the field variables but have continued with the cumbersome notation of Section 9.

Apparently one approximation has consisted in setting $W = 0$ in Eqs. (37–2) and (37–4) (Ref. 4, Chapter II). If this were done consistently, it would be found that the eigenfrequencies are not all real, so that the approximation would completely change the character of the solution. This has escaped notice for a reason that cannot readily be determined. It is apparently connected with implicit approximations (or actual mistakes) that affect Eqs. (37–4), (37–5) and (37–6).[1]

The objective of these approximations is, of course, the simplification of the equations to such an extent that they can easily be solved. If this is once admitted, the question arises: "What terms of the equations are responsible for the mathematical difficulties?" It will appear that the terms proportional to Ω_H in Eqs. (37–2) and (37–4) are the offenders and that the worst difficulties arise from the term $\Omega_H W$ in Eq. (37–2). Moreover, all published treatments of the problem agree in the neglect of this term. This may therefore be called the traditional approximation.

Under many circumstances, it is possible to justify this approximation. This is true whenever the vertical velocity W is much less than the horizontal velocities U and V. In Chapter III, it has been seen that this is true for steady convection; observationally it is true for all large-scale motions on the earth. It cannot be foreseen, without detailed study, whether it will be true under all circumstances. It must none the less be treated as an hypothesis worth considering seriously for technical reasons.

If only the term $\Omega_H W$ were omitted from Eq. (37–1), and the term $\Omega_H U$ were retained in Eq. (37–4), the equations would cease to be self-adjoint. That is, at least some of the eigenfrequencies would become complex, and the nature of the field equations would be completely altered. This indicates that there is a close relation between the two terms in Ω_H, and that if one is neglected, the other must also be neglected, otherwise the approximate equations will surely describe radically-different phenomena than the original equations.

In this extended form, the traditional hypothesis therefore amounts to neglecting the horizontal component, Ω_H, of the Coriolis vector. This extended form of the hypothesis has been used, without explanation, by some earlier writers. If the level surfaces are plane, and the axis of rotation perpendicular to them, it ceases to be an approximation and becomes an idealization; this case will be studied in Chapter IX.

Adopting the traditional approximation, the Eqs. (37–4) and (37–6) yield the first residual equation

$$c\left(\frac{\partial P}{\partial z} + \Gamma P\right) = (N^2 - \omega^2)Q. \tag{7}$$

Equations (37–2) and (37–3) become

$$i\omega U + \Omega_V V = c\,\frac{\partial P}{\partial x},$$

$$i\omega V - \Omega_V U = c\,\frac{\partial P}{\partial y},$$

which can be solved for U and V:

$$\begin{aligned} U &= [c/(\Omega_V^2 - \omega^2)]\left[i\omega\,\frac{\partial P}{\partial x} - \Omega_V\,\frac{\partial P}{\partial y}\right], \\ V &= [c/(\Omega_V^2 - \omega^2)]\left[i\omega\,\frac{\partial P}{\partial y} + \Omega_V\,\frac{\partial P}{\partial x}\right]. \end{aligned} \tag{8}$$

Equations (37–6) and (37–8) can be summarized as

$$\mathbf{U} = [c/(\Omega_V^2 - \omega^2)][i\omega\nabla_H P + \Omega_V \mathbf{k} \times \nabla_H P] + i\omega Q\mathbf{k}. \tag{9}$$

The first term in the bracket is the generalized gradient velocity; the second is the generalized geostrophic; and the last term is the thermobaric.

Substituting Eqs. (37–6) and (37–8) into Eq. (37–5) and noting that Ω_V is a constant while c depends only on z, the second residual equation is obtained as

$$c\left(\frac{\partial Q}{\partial z} - \Gamma Q\right) = P + \frac{c^2}{\omega^2 - \Omega_V^2}\left(\frac{\partial^2 P}{\partial x^2} + \frac{\partial^2 P}{\partial y^2}\right). \tag{10}$$

This differs from Eq. (36–8) only in that ω^2 is replaced by $\omega^2 - \Omega_V^2$.

Separation of variables: two-dimensional wave equation. Once more, the variables in the residual are well separated, and they have solutions which factorize into a function of x, y and a function of z. With the previous notation, the equations to be solved are

$$\begin{aligned} c\left(\frac{\mathrm{d}P_2}{\mathrm{d}z} + \Gamma P_2\right) &= (N^2 - \omega^2)Q_2, \\ c\left(\frac{\mathrm{d}Q_2}{\mathrm{d}z} - \Gamma Q_2\right) &= [1 - \varkappa^2 c^2/(\omega^2 - \Omega_V{}^2)]P_2, \end{aligned} \tag{11}$$

(which differ only trivially from Eqs. (36–12)) and the two-dimensional wave equation, Eq. (36–11).

Calculation without the approximation. These calculations can, of course, also be made without approximation, but the resulting equations are quite elaborate. Equations (37–8) become

$$\begin{aligned} U &= [c/(\Omega_V{}^2 - \omega^2)]\left[i\omega\frac{\partial P}{\partial x} - \Omega_V\frac{\partial P}{\partial y} - \omega^2\Omega_H Q/c\right], \\ V &= [c/(\Omega_V{}^2 - \omega^2)]\left[i\omega\frac{\partial P}{\partial y} + \Omega_V\frac{\partial P}{\partial x} + i\omega\Omega_V\Omega_H Q/c\right]. \end{aligned} \tag{12}$$

Using the abbreviation

$$\mathbf{G} = \nabla_H P + \mathbf{i}(i\omega\Omega_H Q/c), \tag{13}$$

Eq. (37–9) becomes

$$\mathbf{U} = [c/(\Omega_V{}^2 - \omega^2)][i\omega\mathbf{G} + \Omega_V\mathbf{k}\times\mathbf{G}] + i\omega Q\mathbf{k}. \tag{14}$$

The first residual equation, Eq. (37–7), becomes

$$c\left(\frac{\partial P}{\partial z} + \Gamma P\right) + \frac{c\Omega_H}{\omega^2 - \Omega_V{}^2}\left[i\omega\frac{\partial P}{\partial x} - \Omega_V\frac{\partial P}{\partial y}\right] = \left[N^2 - \omega^2\frac{\Omega^2 - \omega^2}{\Omega_V{}^2 - \omega^2}\right]Q, \tag{15}$$

while the second turns into

$$c\left(\frac{\partial Q}{\partial z} - \Gamma Q\right) - \frac{c\Omega_H}{\omega^2 - \Omega_V^2}\left[i\omega\frac{\partial Q}{\partial x} + \Omega_V\frac{\partial Q}{\partial y}\right]$$

$$= P + \frac{c^2}{\omega^2 - \Omega_V^2}\left(\frac{\partial^2 P}{\partial x^2} + \frac{\partial^2 P}{\partial y^2}\right). \quad (16)$$

Separation of variables. The variables in these last two equations are separated but not as symmetrically as in the cases encountered above. The only possibilities for the functions of x, y, are

$$P_1(x, y) = Q_1(x, y) = \exp(i\alpha x + i\beta y), \quad (17)$$

where α and β are constants. The functions $P_2(z)$, $Q_1(z)$ must then satisfy the equations

$$\begin{aligned} c\left(\frac{dP_2}{dz} + \Gamma_e P_2\right) &= \left[N^2 - \omega^2\frac{\Omega^2 - \omega^2}{\Omega_V^2 - \omega^2}\right]Q_2, \\ c\left(\frac{dQ_2}{dz} - \Gamma_e^* Q_2\right) &= \left[1 - \frac{\varkappa^2 c^2}{\omega^2 - \Omega_V^2}\right]P_2, \end{aligned} \quad (18)$$

where

$$\begin{aligned} \Gamma_e &= \Gamma - \frac{\Omega_H}{\omega^2 - \Omega_V^2}(\omega\alpha + i\Omega_V\beta), \\ \varkappa^2 &= \alpha^2 + \beta^2. \end{aligned} \quad (19)$$

Equation (37–18) have much the same form as before, but the fact that Γ_e now has complex values is not a trivial difference.

38. Rotation with spherical level surfaces

When the level surfaces are spherical, the Mercator co-ordinates are again useful because the calculations of the last section then provide clues for those of this one. The vector $\mathbf{\Omega}$ is a constant in space, but its vertical and horizontal components are now functions of the latitude:

$$\mathbf{\Omega} = \Omega_H \boldsymbol{\nu} + \Omega_V \boldsymbol{\zeta}, \quad (1)$$

where

$$\begin{aligned} \Omega_H &= \Omega\cos\phi = \Omega\,\mathrm{sech}\,\mu, \\ \Omega_V &= \Omega\sin\phi = \Omega\tanh\mu, \end{aligned} \quad (2)$$

are now both functions of the horizontal co-ordinate μ. The equations for the eigensolutions—Eqs. (35–1, 2, 3)—are

$$i\omega U_\epsilon + \Omega_V U_\nu - \Omega_H U_\zeta = (c/r)\cosh\mu\,\frac{\partial P}{\partial\lambda}, \tag{3}$$

$$i\omega U_\nu - \Omega_V U_\epsilon = (c/r)\cosh\mu\,\frac{\partial P}{\partial\mu}, \tag{4}$$

$$i\omega U_\zeta + \Omega_H U_\epsilon = c\left(\frac{\partial P}{\partial r} + \Gamma P\right) - N^2 Q, \tag{5}$$

$$i\omega P = (c/r)\cosh\mu\left[\frac{\partial U_\epsilon}{\partial\lambda} + \cosh\mu\,\frac{\partial}{\partial\mu}(U_\nu \operatorname{sech}\mu)\right] + c\left[\frac{1}{r^2}\frac{\partial}{\partial r}(r^2 U_\zeta) - \Gamma U_\zeta\right], \tag{6}$$

$$i\omega Q = U_\zeta. \tag{7}$$

As in Section 36, these equations can be somewhat simplified by replacing r by r_s, $\partial/\partial r$ by $\partial/\partial\chi$. Since this approximation is usually justifiable, it will be adopted in all that follows.

The traditional approximations. Following the plan of the last section, the calculation will be made with the neglect of both terms that involve Ω_H. Then the first residual equation is obtained from Eq. (38–5):

$$c\left(\frac{\partial P}{\partial\chi} + \Gamma P\right) = (N^2 - \omega^2)Q. \tag{8}$$

Equations (38–3) and (38–4) can be solved for U_ϵ and U_ν:

$$U_\epsilon = \frac{c\cosh\mu}{r_s(\Omega_V{}^2 - \omega^2)}\left[i\omega\frac{\partial P}{\partial\lambda} - \Omega_V\frac{\partial P}{\partial\mu}\right],$$
$$U_\nu = \frac{c\cosh\mu}{r_s(\Omega_V{}^2 - \omega^2)}\left[i\omega\frac{\partial P}{\partial\mu} + \Omega_V\frac{\partial P}{\partial\lambda}\right]. \tag{9}$$

When Eqs. (38–7) and (38–9) are substituted into Eq. (38–6), the second residual equation is obtained:

$$c\left(\frac{\partial Q}{\partial\chi} - \Gamma Q\right) = P + \frac{c^2}{\omega^2 r_s{}^2}\,\Delta_\Omega P, \tag{10}$$

where the differential operator Δ_Ω is defined by

$$\Delta_\Omega P = \cosh^2 \mu \left\{ \frac{\partial}{\partial \lambda} \frac{\omega^2}{\omega^2 - \Omega_V{}^2} \left[\frac{\partial P}{\partial \lambda} + i \frac{\Omega_V}{\omega} \frac{\partial P}{\partial \mu} \right] + \frac{\partial}{\partial \mu} \frac{\omega^2}{\omega^2 - \Omega_V{}^2} \left[\frac{\partial P}{\partial \mu} - i \frac{\Omega_V}{\omega} \frac{\partial P}{\partial \lambda} \right] \right\}. \quad (11)$$

Separation of variables: Laplace's tidal equation. The variables in these equations are again well separated, and they have special solutions that factorize as in Eq. (36–23). The factors P_2, Q_2 again satisfy Eqs. (36–12), while the factors P_1, Q_1, which depend only on the Mercator co-ordinates λ, μ, satisfy the equation

$$\Delta_\Omega P_1 + \varkappa^2 r_s{}^2 P_1 = 0, \qquad P_1 = Q_1. \quad (12)$$

Laplace was the first to derive an equation of this form(3) in connection with the theory of tides. It is seen that his equation might well be called "the wave equation on a rotating sphere," were it not for the simplifying assumptions that enter its derivation.

Omission of the traditional approximations. This calculation can also be made without using the traditional approximations. The resulting equations are not only more complicated than those just derived, but the variables are not separated. No study of them has yet been made, and none will be attempted in this work.

39. Complex vectors and the hodograph

In all calculations with eigensolutions, free use will be made of complex numbers. In general, the field $P = P_r + iP_i$, where P_r and P_i are real functions of the co-ordinates. In addition, the time dependence is given by the factor $\exp(-i\omega t)$; the negative sign in this exponent is conventional in theoretical physics and contrary to the convention in electronics. The complete expression for the field P is thus

$$P = (P_r + iP_i) \exp(-i\omega t) = [P_r \cos \omega t + P_i \sin \omega t] + i[-P_r \sin \omega t + P_i \cos \omega t]. \quad (1)$$

The same applies to the vector field $\mathbf{U}$:

$$\mathbf{U} = (\mathbf{U}_r + i\mathbf{U}_i) \exp(-i\omega t) = [\mathbf{U}_r \cos \omega t + \mathbf{U}_i \sin \omega t] + i[-\mathbf{U}_r \sin \omega t + \mathbf{U}_i \cos \omega t]. \quad (2)$$

It will be noted that the imaginary parts of both these expressions can be obtained from the real parts by substituting $t + \pi/2\omega$ for t; it therefore suffices to study the real parts.

In the case of the first equation, P can be represented graphically by a two-dimensional rotating vector; to do the same for the second equation would require six dimensions. It is therefore more convenient to study only the real part,

$$\mathscr{R}\mathbf{U} = \mathbf{U}_r \cos \omega t + \mathbf{U}_i \sin \omega t, \tag{3}$$

which can be represented by a moving three-dimensional vector. If the origin of this vector is held fixed, its tip describes an ellipse, this term including the limiting cases of a circle and a straight line segment.

The ellipse lies in the plane of the two vectors $\mathbf{U}_r$ and $\mathbf{U}_i$, which are conjugate radii. It is called the hodograph; the eigensolutions are said to be elliptically-, circularly-, or linearly-polarized, depending upon the character of the hodograph. Since the vectors $\mathbf{U}_r$ and $\mathbf{U}_i$ are functions of position in space, the shape of the hodograph may change from point to point, and the polarization will vary correspondingly. In many cases, however, the polarization will have the same type at all points in space.

No rotation. Equations (26–8, 9) and Eqs. (36–18, 19) show that the fields P and Q can be supposed to be real in this case. Equation (36–6) shows that then

$$\begin{aligned} \mathbf{U}_r &= 0, \\ \mathbf{U}_i &= -(c/\omega)\nabla_H P + \omega Q\mathbf{k}, \end{aligned} \tag{4}$$

and the eigensolutions are all linearly-polarized. The direction and magnitude of $\mathbf{U}_i$ will vary from point to point; this variation cannot be discussed without more detailed knowledge of the solutions of the residual equations.

Rotation: traditional approximation. In this case, Eq. (37–9) shows that

$$\begin{aligned} \mathbf{U}_r &= \frac{c\Omega_V}{\Omega_V{}^2 - \omega^2}\, \mathbf{k} \times \nabla_H P, \\ \mathbf{U}_i &= \frac{c\omega}{\Omega_V{}^2 - \omega^2}\, \nabla_H P + \omega Q\mathbf{k}. \end{aligned} \tag{5}$$

Since these vectors are mutually perpendicular, they are principal radii of the elliptical hodograph, and $\mathbf{U}_r$ is horizontal. If $\Omega_V > \omega$, and Q is not too large, the major axis will be horizontal and have the direction of the geostrophic velocity. On the other hand, if $\omega > \Omega_V$, this will usually be the minor axis of the ellipse, and the gradient and thermobaric velocities will determine its major axis.

Rotation, without approximation. This calculation has been made, above, only for the case of plane level surfaces, but the following remarks probably apply also to curved level surfaces. Equations (37–18) show that P and Q will themselves be complex. The same will apply to the vector $\mathbf{G}$, Eq. (37–13), so that

$$\begin{aligned} \mathbf{G}_r &= \nabla_H P_r - \mathbf{i}\Omega_H \omega Q_i/c, \\ \mathbf{G}_i &= \nabla_H P_i + \mathbf{i}\Omega_H \omega Q_r/c. \end{aligned} \tag{6}$$

The vector $\mathbf{G}$ will therefore itself describe an ellipse in the horizontal plane, whose shape is not readily determined. The hodograph can be determined from Eq. (37–14), but about the most that can be said is that it will usually be an ellipse.

References for Chapter VII

1. C. Eckart and H. G. Ferris, Equations of motion of the ocean and atmosphere, *Revs. Mod. Phys.* **28**, 48 (1956).
2. S. S. Hough, An application of harmonic analysis to the dynamical theory of tides, Part I, *Phil. Trans. (London)* **A189**, 201 (1897).
3. P. S. Laplace, *Méchanique Céleste*, 4 vols. (Trans. by N. Bowditch), Boston, 1832. The relevant section is Part I, Book IV, Section 3, p. 543.
4. M. V. Wilkes, *Oscillations of the Earth's Atmosphere*, 74 pp., Cambridge Univ. Press, 1949.

CHAPTER VIII

THE ISOTHERMAL ATMOSPHERE: PLANE LEVEL SURFACES WITHOUT ROTATION

40. Introduction

THE next two chapters will be devoted to calculations concerning the isothermal atmosphere. Although the Earth's atmosphere is far from isothermal, this case is of illustrative value because the coefficients, c, N and Γ, of the field equations are all constants, independent of the altitude This simplifies the calculation very markedly and permits concentration on certain aspects of the problem without the distraction of analytic difficulties. In Chapter X the oceanic case of constant coefficients will be examined. By then, certain general methods and principles will have been found, and the way will have been prepared for the discussion of more elaborate problems.

In the present chapter, the case of no rotation will be treated in detail, and even with some repetition of matters briefly discussed in the last chapter. If the level surfaces are plane, the horizontal velocity consists only of the gradient component, and the thermobaric velocity is vertical. The further discussion can start with Eqs. (36–8, 9), which are

$$c\left(\frac{\partial P}{\partial z} + \Gamma P\right) = (N^2 - \omega^2)Q,$$
$$c\left(\frac{\partial Q}{\partial z} - \Gamma Q\right) = P + \frac{c^2}{\omega^2}\left(\frac{\partial^2 P}{\partial x^2} + \frac{\partial^2 P}{\partial y^2}\right). \tag{1}$$

The planet's surface being treated as the plane $z = 0$, the only boundary condition requires that the vertical velocity vanish there. This is equivalent to

$$Q = 0 \qquad \text{at} \qquad z = 0, \tag{2}$$

unless ω is zero. This alternative will not be considered further.

The first pair of equations can be reduced to a pair of ordinary differential equations by considering only those eigensolutions for which P and Q are proportional to

$$\exp\,[i(\alpha x + \beta y)],$$

α and β being constants; the factors of proportionality are to be functions of z only. Using P and Q, now, to denote these factors, Eqs. (40–1) become

$$\begin{aligned} c\left(\frac{\mathrm{d}P}{\mathrm{d}z} + \Gamma P\right) &= (N^2 - \omega^2)Q, \\ c\left(\frac{\mathrm{d}Q}{\mathrm{d}z} - \Gamma Q\right) &= (1 - \varkappa^2 c^2/\omega^2)P, \end{aligned} \tag{3}$$

where

$$\varkappa^2 = \alpha^2 + \beta^2.$$

These are to be solved subject to the boundary condition of Eq. (40–2). The constant $\varkappa$ will be of importance throughout the following; it will be called the horizontal wave number, and $2\pi/\varkappa$, the horizontal wavelength. The vector $\alpha\mathbf{i} + \beta\mathbf{j}$ is the horizontal propagation vector. The ratio $\omega/\varkappa$ has the dimensions of a velocity and will be called the trace velocity of the waves.

41. Lamb's waves

Lamb (Ref. 2, p. 548) has found some eigensolutions for this problem. The boundary condition, Eq. (40–2), will certainly be satisfied if $Q = 0$ for all z, so that the waves are locally isentropic everywhere. Equations (40–3) then yield

$$P = \exp\,(-\Gamma z) \tag{1}$$

$$\varkappa^2 c^2 = \omega^2; \tag{2}$$

consequently the trace velocity is that of sound. This solution will be regular (Section 20) if $\Gamma \geqslant 0$, and it has been seen (Section 25) that this is the case for an isothermal atmosphere:

$$\Gamma = g(2 - \gamma)/2\gamma R\theta_0.$$

Recalling that for an isothermal atmosphere (Section 22),

$$\rho_0 c_0 = \rho_s c_s \exp\,(-gz/R\theta_0),$$

Eq. (41–1) may be written

$$p_1 = (\rho_s c_s)^{\frac{1}{2}} \exp[-gz/\gamma R\theta_0).$$

Since $Q = 0$, the first-order entropy is zero: the changes in the pressure and density do not change the local value of the entropy. For the same reason, the thermobaric component of the velocity vanishes, and the wave is linearly-polarized in the direction of propagation:

$$\mathbf{U} = [(\mathbf{i}\alpha + \mathbf{j}\beta)/\varkappa]P; \tag{3}$$

the kinetic energy $|U|^2/2c$ is equal to the elastic energy $|P|^2/2c$, while the thermobaric energy is zero. This result may also be written

$$|\mathbf{U}_1| = |p_1|/\rho c$$

so that the characteristic impedance is ρc. All of this, except the variation of amplitude with altitude, is strongly reminiscent of a sound wave in a homogeneous fluid. The variation of density with altitude has only two effects: it causes the amplitude to diminish according to Eq. (41–1) and prevents the propagation of waves in any except horizontal directions.

Strictly speaking, Lamb's waves are unlimited eigensolutions. However, if one calculates only the energy contained in a column of the atmosphere, 1 cm^2 in cross-section, but extending from $z = 0$ to $z = \infty$, this will be finite. In a modified sense, Lamb's waves are therefore limited. They are not the only eigensolutions of the problem, but all others are unlimited even in this modified sense. This may be the reason why Lamb did not consider the others.

42. Other eigensolutions; simple waves

The boundary condition, Eq. (40–2), will be satisfied if Q is proportional to

$$\sin \gamma z \exp[i(\alpha\varkappa + \beta y)],$$

γ being a third constant. Such solutions will be regular only if γ is real—a fact that will be used recurrently below. The second of Eqs. (40–1) then shows that P will be proportional to

$$[\gamma \cos \gamma z - \Gamma \sin \gamma z] \exp[i(\alpha x + \beta y)].$$

The algebra will be much simplified if these functions are written

$$(1/2i)[\exp \{i(\alpha x + \beta y + \gamma z)\} - \exp \{i(\alpha x + \beta y - \gamma z)\}],$$

and

$$(1/2i)[-(i\gamma - \Gamma) \exp \{i(\alpha x + \beta y + \gamma z)\} + (i\gamma + \Gamma) \exp \{i(\alpha x + \beta y - \gamma z)\}].$$

respectively. The second terms of these functions represent waves incident upon the ground in the direction α, β, $-\gamma$; the first terms represent waves reflected by the ground into the direction α, β, γ. Incident and reflected waves, taken separately, must satisfy the differential equations but only their difference satisfies the boundary condition. It is therefore permissible to postpone (or, for some purposes, even to omit) the boundary condition and to examine the two terms separately. Each term may be called a simple wave; simple waves are not eigensolutions by themselves but must be so combined as to satisfy the boundary condition. The vector $\alpha\mathbf{i} + \beta\mathbf{j} \pm \gamma\mathbf{k}$ is the propagation vector, and its magnitude,

$$k = (\alpha^2 + \beta^2 + \gamma^2)^{\frac{1}{2}}$$

is the wave number; the wave-length is $2\pi/k$.

For simple waves, Eqs. (40–1) become

$$\begin{aligned} c(\Gamma + i\gamma)P - (N^2 - \omega^2)Q &= 0, \\ (1 - \varkappa^2 c^2/\omega^2)P + c(\Gamma - i\gamma)Q &= 0. \end{aligned} \tag{1}$$

In order that these equations may have other than the trivial solution, their determinant must vanish:

$$c^2(\Gamma^2 + \gamma^2) + (N^2 - \omega^2)(1 - \varkappa^2 c^2/\omega^2) = 0. \tag{2}$$

Various authors give various names to equations of this sort; it will be less confusing to adopt the name "propagation equation" for the second equation than to use the more common term "characteristic equation."

If ω and $\varkappa$ are given, the propagation equation determines γ^2. The value so determined may be negative: since γ must be real, such values of ω and $\varkappa$ do not lead to eigensolutions. This situation is conveniently exhibited graphically as in Fig. 16; Eq. (42–2) is equivalent to

$$\gamma^2 c^2 = (\omega^2 - N_1^2) + \varkappa^2 c^2 (N^2 - \omega^2)/\omega^2, \tag{3}$$

where

$$N_1^2 = N^2 + \Gamma^2 c^2. \tag{4}$$

The curve $\gamma^2 = 0$ consists of two branches, one of which intersects the ω-axis at $\omega = N_1$ and becomes asymptotic to the line $\omega = \varkappa c$. The other branch passes through the origin and becomes asymptotic to the horizontal line $\omega = N$. At the origin, it has the slope $(N/N_1)c$, which

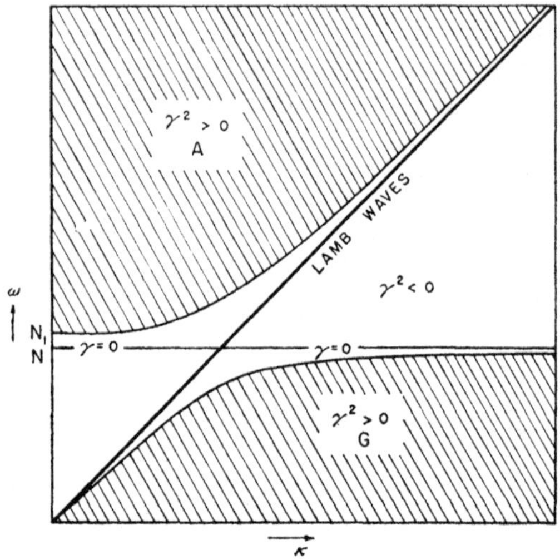

FIG. 16. Diagnostic, or $\varkappa$, ω diagram for an isothermal atmosphere.

is to be compared with that of the curve for Lamb's waves, which is c. These branches divide the first quadrant of the $(\omega,\varkappa)$ plane into three areas: $\gamma^2 > 0$ in those identified as A and G, while $\gamma^2 < 0$ in the third. In this third area, only the points on the line $\omega = \varkappa c$ lead to eigensolutions (i.e. to Lamb's waves). In the areas A and G, every point leads to an eigensolution.

Diagrams of this sort will be almost essential in discussing these problems. For the present problems, they are as important as the energy level diagrams in atomic physics, which have some points of mathematical similarity. This $(\varkappa,\omega)$ diagram will be called the diagnostic diagram.

Equations (25–1, 2, 3) show that for an isothermal gas

$$\Gamma^2 c^2/N^2 = (2 - \gamma)^2/4(\gamma - 1),$$

γ being here the ratio of specific heats. (This ambiguity of notation will not be serious.) For air, $\gamma = 1{\cdot}40$, so that $N_1 = 1{\cdot}11N$.

43. The propagation surface; phase velocity

A second kind of diagram is suggested by writing the propagation equation, Eq. (42–2), in the form

$$\frac{(\alpha^2 + \beta^2)(\omega^2 - N^2)}{\omega^2(\omega^2 - N_1^2)} + \frac{\gamma^2}{(\omega^2 - N_1^2)} = \frac{1}{c^2}. \tag{1}$$

If ω is given a fixed value, this represents a quadric surface in the (α,β,γ) space, which is usually called the propagation surface (Ref. 3, p. 292).

When $\omega^2 > N_1^2$ (area A of the diagnostic), the propagation surface is an oblate ellipsoid of revolution, whose horizontal semi-axes have the length

$$(\omega/c)[(\omega^2 - N_1^2)/(\omega^2 - N^2)]^{\frac{1}{2}}$$

and whose vertical semi-axis is

$$(\omega^2 - N_1^2)^{\frac{1}{2}}/c.$$

The propagation vector, $\mathbf{k}$, is the radius vector from the origin to a point on the surface. As its direction varies, ω being held constant, the wave number k varies in a manner that will be clear from Fig. 17. The phase of the simple waves will be constant on the plane

$$\alpha x + \beta y + \gamma z = \text{const.},$$

which is perpendicular to $\mathbf{k}$, and whose trace is shown by the dotted line of Fig. 17.

When $N^2 < \omega^2 < N_1^2$, the propagation surface is imaginary (cf. the unshaded area of Fig. 16).

When $0 < \omega^2 < N^2$ (area G of Fig. 16), the propagation surface is an hyperboloid of revolution, as shown in Fig. 17. Its semi-axes are

$$(\omega/c)[(N_1^2 - \omega^2)/(N^2 - \omega^2)]^{\frac{1}{2}}$$

and

$$[N_1^2 - \omega^2]^{\frac{1}{2}}/c.$$

The asymptotic cone of the hyperboloid has the positive γ-axis as its axis, and the half-angle θ, given by

$$\tan \theta = \omega/[N^2 - \omega^2]^{\frac{1}{2}}. \tag{2}$$

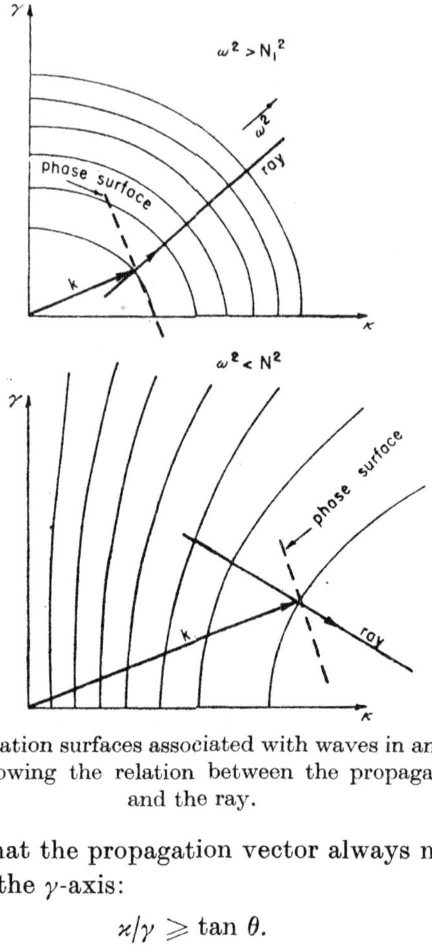

FIG. 17. Propagation surfaces associated with waves in an isothermal atmosphere, showing the relation between the propagation vector and the ray.

Figure 17 shows that the propagation vector always makes a greater angle than θ with the γ-axis:

$$\varkappa/\gamma \geqslant \tan \theta.$$

In particular, no waves of the type G have a vertical propagation vector. Those whose propagation vector falls on the asymptotic cone have an infinite wave number and zero wave-length.

The phase velocity of the simple waves is

$$\begin{aligned} c_{\text{ph}} &= \omega/k \\ &= c\left\{\frac{\omega^2 - N^2(\varkappa^2/k^2)}{\omega^2 - N_1^2}\right\}^{\frac{1}{2}}. \end{aligned} \tag{3}$$

For $\omega > N$, $c_{\text{ph}} > c$, but as $\omega \to \infty$, $c_{\text{ph}} \to c$. For $\omega < N$, $c_{\text{ph}} < c$, and approaches $c(\varkappa/k)(N/N_1)$ as $\omega \to 0$. Since the phase velocity depends on both frequency and direction, the propagation is both dispersive and anisotropic.

44. Rays and group velocity

In order to arrive at a preliminary definition of the rays and group velocity, it is convenient to suppose that the propagation equation, Eq. (42–2), has been solved for ω as a function of α, β and γ. Since that equation is of second degree in ω^2, it will have four roots

$$\omega = \pm H_1(\alpha, \beta, \gamma), \qquad \omega = \pm H_2(\alpha, \beta, \gamma).$$

Let H be any one of these, and consider two superposed waves, of equal amplitude but slightly differing propagation vectors, so that, for example

$$Q = \sin\,[\alpha' x + \beta' y + \gamma' z - H(\alpha', \beta', \gamma')t] + \sin\,[\alpha'' x + \beta'' y + \gamma'' z - H(\alpha'', \beta'', \gamma'')t].$$

If $\alpha' = \alpha + \delta\alpha$, $\alpha'' = \alpha - \delta\alpha$, etc., this becomes approximately

$$Q = 2 \sin\,(\alpha x + \beta y + \gamma z - Ht) \cos\left[\left(x - t\frac{\partial H}{\partial \alpha}\right)\delta\alpha + \left(y - t\frac{\partial H}{\partial \beta}\right)\delta\beta + \left(z - t\frac{\partial H}{\partial \gamma}\right)\delta\gamma\right].$$

This result may be described as a modulated carrier wave, the carrier having the frequency H, and the propagation vector (α,β,γ). The modulation is also a wave, whose propagation vector is $(\delta\alpha,\delta\beta,\delta\gamma)$, and whose frequency is

$$\delta\omega = \frac{\partial H}{\partial \alpha}\,\delta\alpha + \frac{\partial H}{\partial \beta}\,\delta\beta + \frac{\partial H}{\partial \gamma}\,\delta\gamma.$$

Its direction of propagation is not the same as that of the carrier, but entirely independent of the carrier. Its frequency depends on all six variables α, β, γ, $\delta\alpha$, $\delta\beta$, $\delta\gamma$, and will be low.

However, regardless of the values of $\delta\alpha$, $\delta\beta$, $\delta\gamma$, the modulation will be constant at any point which moves according to the equations

$$x - t\frac{\partial H}{\partial \alpha} = \text{const.} = x_0,$$

$$y - t\frac{\partial H}{\partial \beta} = \text{const.} = y_0, \qquad (1)$$

$$z - t\frac{\partial H}{\partial \gamma} = \text{const.} = z_0.$$

These straight lines have a direction that is independent of $\delta\alpha$, $\delta\beta$, $\delta\gamma$, and the moving point has the velocity, c_{gr}, defined by

$$c_{gr} = \left[\left(\frac{\partial H}{\partial \alpha}\right)^2 + \left(\frac{\partial H}{\partial \beta}\right)^2 + \left(\frac{\partial H}{\partial \gamma}\right)^2\right]^{\frac{1}{2}}. \qquad (2)$$

The straight lines are the rays, and c_{gr} is the group velocity. The functions $\pm H_1$, and $\pm H_2$ are the Hamiltonians of the rays.[(1)]

It is at once seen that the rays do not have the same direction as the propagation vector but are perpendicular to the surface $H = \omega =$ constant. That is, they are perpendicular to the propagation surface which passes through the end of propagation vector, as shown on Fig. 17.

The propagation equation may be written

$$W(\alpha, \beta, \gamma, \omega) = 0, \qquad (3)$$

provided that

$$W = c^2(\alpha^2 + \beta^2)(\omega^2 - N^2) + \gamma^2 c^2 \omega^2 - \omega^2(\omega^2 - N_1^2). \qquad (4)$$

Clearly

$$W = -(\omega^2 - H_1^2)(\omega^2 - H_2^2). \qquad (5)$$

Hence, no matter which root H may represent,

$$\frac{\partial H}{\partial \alpha} = -\frac{\partial W}{\partial \alpha}\bigg/\frac{\partial W}{\partial \omega}, \quad \text{etc.,}$$

provided only that α, β, γ, ω satisfy Eq. (44–3). The equation of the rays, Eq. (44–1), may therefore be written

$$(x - x_0)\frac{\partial W}{\partial \omega} + t\frac{\partial W}{\partial \alpha} = 0,$$
$$(y - y_0)\frac{\partial W}{\partial \omega} + t\frac{\partial W}{\partial \beta} = 0, \tag{6}$$
$$(z - z_0)\frac{\partial W}{\partial \omega} + t\frac{\partial W}{\partial \gamma} = 0.$$

These equations are rational, while Eqs. (44–1) contain radicals; W may be called the rational Hamiltonian of the rays (Ref. 1, p. 412; Ref. 3, p. 286).

In the present case, Eqs. (44–4) and (44–6) yield the explicit equations for the rays, in the form

$$x - x_0 = c_x t,$$
$$y - y_0 = c_y t, \tag{7}$$
$$z - z_0 = c_z t,$$

where

$$c_x = (\alpha c^2/\omega)R,$$
$$c_y = (\beta c^2/\omega)R, \tag{8}$$
$$c_z = (\gamma c^2/\omega)[\omega^2/(\omega^2 - N^2)]R,$$

and

$$R = (\omega^2 - N^2)/(2\omega^2 - N_1^2 - k^2c^2). \tag{9}$$

The algebraic identities

$$\frac{1}{R} = 1 + (\gamma^2 + \Gamma^2)c^2\frac{N^2}{(\omega^2 - N^2)^2} - \frac{W}{(\omega^2 - N^2)^2}, \tag{10}$$

$$= 1 + \frac{N^2}{\omega^2}\frac{\omega^2 - \varkappa^2c^2}{\omega^2 - N^2} - \frac{W}{\omega^2(\omega^2 - N^2)}, \tag{11}$$

are readily verified; since $W = 0$, they provide alternative formulae for R. Equation (44–9) shows that R is independent of the direction of propagation; Eq. (44–10) shows that for $W = 0$ and real values of

γ, $R \leqslant 1$; Eq. (44–11) is convenient for plotting the curves $R =$ constant on the $(\omega,\varkappa)$ plane as has been done in Fig. 18.

It will be seen in Section 45 that R is the characteristic impedance of the waves, measured in units of ρc. Throughout the greater part of area A, its value is nearly unity, which is its value in classical acoustics. It diminishes toward (but never reaches) the value zero

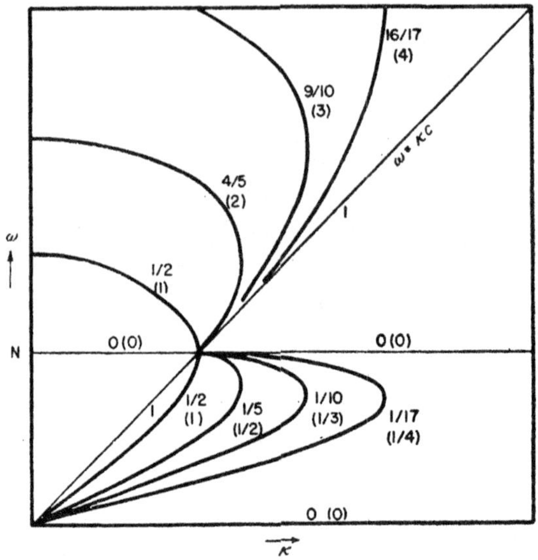

FIG. 18. The dimensionless impedances, R and S, as functions of $\varkappa$ and ω. Numbers in parentheses are values of S; the others are values of R.

in the lowest portions of area A. Its value is nearly zero throughout the greater part of area G but increases, without reaching the value unity, in the left part of the area.

45. The pressure–entropy impedance

Equation (42–1) shows that in a simple wave

$$Q = Pc(i\gamma + \Gamma)/(N^2 - \omega^2). \tag{1}$$

In general both P and Q are complex numbers. The equation above shows that their phase angles differ by $\tan^{-1}(\gamma/\Gamma)$. In waves of type G, the entropy leads the pressure (in time) by an angle that is nearly 90° except near the curved boundary of the area, where $\gamma = 0$, and the two are in phase. In waves of type A, the entropy lags behind the pressure by an angle that is nearly 90° except near the boundary $\gamma = 0$, where it is 180°.

The amplitudes of the variation of the two fields are proportional:

$$N|Q| = |P|/S, \tag{2}$$

where $S > 0$ and

$$S^2 = (N^2 - \omega^2)^2/[N^2c^2(\gamma^2 + \Gamma^2)],$$

$$= (\omega^2 - N^2)/[N^2(1 - \varkappa^2c^2/\omega^2)],$$

because of the propagation equation, Eq. (42–2). Comparison of this last equation with Eq. (44–11) shows that

$$S^2 = R/(1 - R). \tag{3}$$

The values of S are shown on Fig. 18 by the numbers in parentheses.

For waves of type A, S is usually very large, so that the changes in the local thermodynamic state are nearly isentropic. This implies that the elastic energy (cf. Sections 21, 22) will be much greater than the thermobaric, except in the lower part of the area A. In classical acoustics the thermobaric energy is assumed to be exactly zero; consequently the waves of type A may be identified with sound waves, as modified by the action of gravity. This identification is substantiated by their phase velocity (Section 43) and will receive further confirmation below.

For waves of type G, S is usually very small, and the local thermodynamic changes occur almost isopiestically. The thermobaric energy is much greater than the elastic, and these waves may be called internal gravity waves, although the adjective "thermobaric" might be more accurately descriptive.

46. The flow and partition of energy in simple waves

In simple waves, the relation between the velocity fields and the pressure field is given by (cf. Eqs. 36–1, 2, 3)

$$U_x = (\alpha c/\omega)P,$$

$$U_y = (\beta c/\omega)P, \tag{1}$$

$$U_z = [(\gamma - i\Gamma)\omega c/(\omega^2 - N^2)]P.$$

The fluid therefore moves neither in the direction of the propagation vector nor in that of the group velocity; except that in the high-frequency limit, all three directions coincide.

The energy density is, by Eq. (33–6):

$$E = (1/2c)(|\mathbf{U}|^2 + |P|^2 + N^2|Q|^2)$$
$$= (1/2c)[(1 + \varkappa^2 c^2/\omega^2) + (\gamma^2 + \Gamma^2)c^2(N^2 + \omega^2)/(N^2 - \omega^2)^2]|P|^2, \tag{2}$$

and the energy flow is, by Eq. (33–7)

$$\begin{aligned} J_x &= \tfrac{1}{2}[PU_x^* + P^*U_x] \\ &= (\alpha c/\omega)|P|^2, \\ J_y &= (\beta c/\omega)|P|^2, \\ J_z &= [\gamma c\omega/(\omega^2 - N^2)]|P|^2. \end{aligned} \tag{3}$$

On comparing this last equation with Eq. (44–8) it is seen that

$$\begin{aligned} J_x &= |P|^2(c_x/Rc), \\ J_y &= |P|^2(c_y/Rc), \\ J_z &= |P|^2(c_z/Rc). \end{aligned} \tag{4}$$

so that the rays are the streamlines of the energy flow. This is an alternative and more fundamental definition of the rays.

Moreover, if Eq. (46–2) is written $E = A|P|^2/c$, the identity

$$2\omega^2(\omega^2 - N^2)^2(A - 1/R) = (3\omega^2 - N^2)W, \tag{5}$$

is readily verified. Since $W = 0$, it follows that Eq. (46–2) is equivalent to

$$E = |P|^2/Rc. \tag{6}$$

Therefore the flow of energy along the rays occurs with the group velocity, since Eqs. (46–4) and (46–5) combine into

$$J_x = Ec_x, \quad \text{etc.}$$

Equation (46–6) may also be written

$$E = |p_1|^2/R\rho c^2. \tag{7}$$

Hence $R\rho c$ is the characteristic impedance of the waves, in c.g.s.

units. The definitions of R and S may also be used to show that the elastic energy density is

$$(1/2c)|P|^2 = \tfrac{1}{2}RE, \tag{8}$$

while the thermobaric is

$$(N^2/2c)|Q|^2 = \tfrac{1}{2}(1 - R)E. \tag{9}$$

By difference, the kinetic energy density is

$$(1/2c)|U|^2 = \tfrac{1}{2}E. \tag{10}$$

47. The eigensolutions

The phase diagram. The simple waves do not satisfy the boundary condition $Q = 0$ at $z = 0$; the construction of the eigen-solutions, which do satisfy this condition, has been described in Section 42. Omitting exponential factors, they are

$$\begin{aligned} P &= P_0 \sin (\psi - \gamma z), \\ NQ &= \pm(P_0/S) \sin \gamma z, \end{aligned} \tag{1}$$

where

$$P_0 > 0, \quad \tan \psi = \gamma/\Gamma, \quad 0 \leqslant \psi \leqslant \pi/2, \tag{2}$$

and P_0 is constant. The double sign arises from the definition $S > 0$; the positive sign is to be used when $\omega^2 > N^2$; the negative, when $\omega^2 < N^2$. The quantities P and NQ have the same units and may be plotted as rectangular co-ordinates of a point—the phase point (cf. Section 7). As the altitude z varies, the phase point describes an ellipse, whose construction is shown on Fig. 19, where $OA = P_0$, $OB = P_0/S$. As z increases, the phase point describes this ellipse in the positive sense for type A waves ($\omega^2 > N^2$), and in the negative for type G waves ($\omega^2 < N^2$). It will appear below that this sense of "motion" of the phase point is of fundamental importance in classifying the eigensolutions.

The eigensolutions may be described as waves whose amplitude varies with altitude as in Eqs. (47–1), that are propagated horizontally with the trace velocity $\omega/\varkappa$. Reference to the diagnostic shows that the trace velocity is greater than c for type A, less than c for type G waves.

For waves of type A, with $\omega \gg N$, Fig. 18 shows that $S \gg 1$, so that $NQ \ll P$ and the ellipse of the phase-diagram is much elongated in the P-direction. In the limit $S \to \infty$, these waves are therefore isentropic and have all the properties of ordinary sound waves.

These properties are modified as $\omega \to N$ and $S \to 0$; the ellipse then elongates in the NQ direction, and the waves are nearly isopiestic. This is also characteristic of the type G waves, for which S is, in general, small. Toward the left of area G, however, S becomes large, and these waves become isentropic and somewhat similar to ordinary sound.

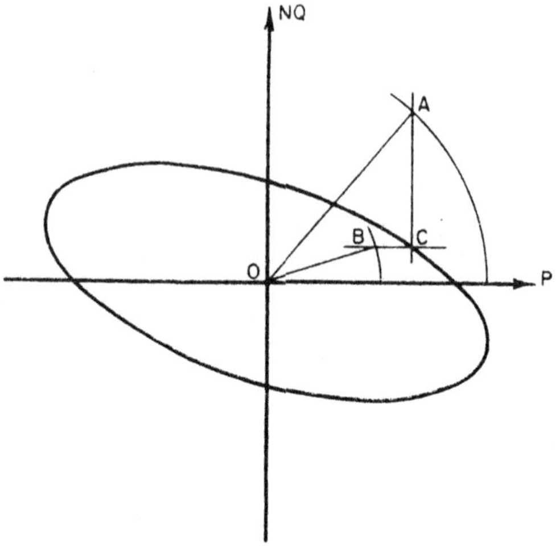

FIG. 19. Phase diagram for the eigensolutions of an isothermal atmosphere. As the altitude z increases, the point C traces the ellipse. The motion of C is clockwise for sound waves, counter-clockwise for gravity waves.

When S is neither very large nor very small, the character of the waves varies with altitude, being isentropic at $\gamma z = n\pi + \psi$ and isopiestic at $\gamma z = N\pi$. At special intermediate altitudes, they become isothermal or isopyknic; in general, they cannot be classified in any of these four categories. This thermodynamic generality of the eigensolutions is a fundamental physical fact about the isothermal atmosphere; all too often, it is assumed that its changes of state occur isentropically or isothermally.

The hodograph. The hodograph of the waves is described by the formulae

$$\begin{aligned} U_H &= P_0(\varkappa c/\omega) \sin (\gamma z - \psi), \\ U_V &= \pm i P_0(\omega/NS) \sin \gamma z, \end{aligned} \tag{3}$$

U_H and U_V being the horizontal and vertical velocity fields. The hodograph always lies in the vertical plane containing the propagation vector and is an ellipse with axis ratio

$$\left|\frac{U_V}{U_H}\right| = \frac{\omega^2}{\varkappa c N S}\left|\frac{\sin \gamma z}{\sin(\gamma z - \psi)}\right|.$$

At the altitudes $\gamma z = n\pi$, it becomes a horizontal straight line; while at $\gamma z = n\pi + \psi$, it is a vertical straight line. The trigonometric

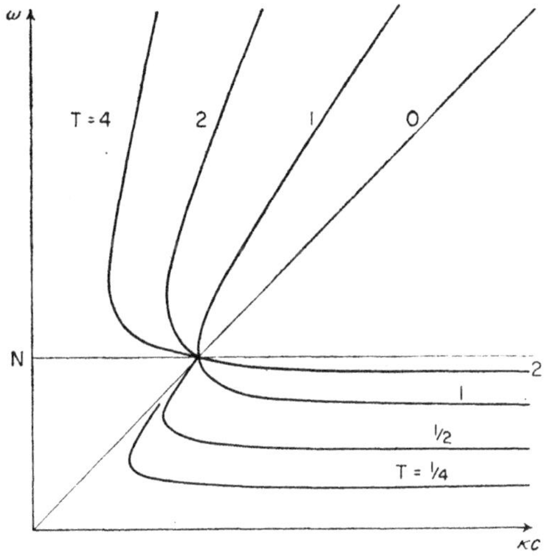

FIG. 20. The ratio, T, as a function of $\varkappa$ and ω.

factor will be of the order of magnitude unity for all except quite narrow ranges of z. The axis ratio will generally have the order of magnitude T, where

$$\begin{aligned} T^2 &= \left(\frac{\omega}{\varkappa c}\right)^2\left(\frac{\omega}{N}\right)^2\frac{1}{S^2} \\ &= \frac{\omega^2(\omega^2 - \varkappa^2c^2)}{\varkappa^2c^2(\omega^2 - N^2)}. \end{aligned} \tag{4}$$

Figure 20 shows the contours of constant T in the diagnostic. For small values of ω/N (waves of type G) the approximation

$$T = \omega/N$$

is valid, provided that $\varkappa c/N$ is not too small.

For small values of ω/N, the vertical velocity will thus be small compared to the horizontal. It will be seen that this is of importance in connection with the traditional approximation discussed in Section 37. For waves of type A, the approximation

$$T = \gamma/\varkappa$$

is valid, except near the curved boundary of area A.

48. The gravity waves and the fluctuating wind

Thus far, the emphasis has been on the deductive aspects of the theory. It remains to identify the various types of solutions with known atmospheric phenomena. This is simple in the case of solutions of type A: they are clearly sound waves. The ordinary theory of acoustics differs from the theory developed here only in one respect: the gravitational effects arising from stratification are usually neglected. It has been seen that these effects are very small for frequencies much greater than N. Since it may be supposed that N is of the order of 1 rad/min, it follows that these gravitational phenomena become appreciable only for sound whose wave-length is more than 50 km. More striking is the fact that there are no sound waves with frequencies less than N. Instead, the Lamb waves more or less take over the role of sound waves at lower frequencies.

The identification of the gravity waves is, perhaps, less obvious. Those of short wave-length—large values of $\varkappa$—have one outstanding characteristic that distinguishes them from sound waves. In the latter, the ratio of the particle velocity to the pressure amplitude is very small—of the order of magnitude $1/\rho c$. In the gravity waves, this ratio becomes much larger and approaches infinity for short wave-lengths. This is also a characteristic of the fluctuations in wind velocity that occur without marked pressure fluctuations. One may therefore make a tentative identification of the gravity waves with the fluctuating component of the wind.

However, this identification is not without some inherent difficulties. It must be recalled that the atmosphere to which the present calculations apply is highly idealized, and that the calculations are themselves approximate. In the present context, the most significant idealization concerns the lower boundary of the fluid, which has been assumed to be a level plane. The Earth's actual surface is far from this—it consists of rocks, trees, hills and mountains. As the air flows around the smaller of these obstacles, turbulent vortices are formed in a manner that is well known from experiments in wind tunnels. These contribute much to the more rapid, localized fluctuation of air speed. They undoubtedly are major phenomena in the lowest layers of the atmosphere. With the idealized boundary, none of these phenomena are included in the calculation, and it is probable that even if the boundary could be treated in a less idealized way, the other approximations would prevent an adequate mathematical description of these phenomena from being obtained.

Consequently, the identification can be made only when the space and time scales of the phenomena are large enough and the distance from the boundary great enough so that its actual departure from an ideal plane is of no great importance.

But, at this stage, another shortcoming of the present calculation makes itself felt: the Earth's atmosphere is far from isothermal. It will be seen in Chapter XII that the best evidence for this identification of the gravity waves with the fluctuating wind is derived from observations made when there is a marked temperature inversion in the lower kilometre of the atmosphere. Under these conditions, it is possible to obtain almost quantitative agreement between theory and observation.

References for Chapter VIII

1. C. Eckart, Approximate solution of one-dimensional wave equations, *Revs. Mod. Phys.* **20**, 399 (1948).
2. H. Lamb, *Hydrodynamics*, Cambridge Univ. Press, 6th ed., 728 pp. 1932.
3. H. Poincaré, *Leçons sur la Théorie Mathématique de la Lumière*, Carré, Paris, 1889.

CHAPTER IX

THE ISOTHERMAL ATMOSPHERE: PLANE LEVEL SURFACES WITH ROTATION

49. Vertical axis of rotation

If the level surfaces are plane and the atmosphere rotates about a vertical axis, the considerations of the last chapter are complicated only slightly, and the same notation can be used. The velocity is conveniently separated into three terms:

$$\mathbf{U} = \mathbf{U}_1 + \mathbf{U}_2 + \mathbf{U}_3,$$

where, in the case of simple waves:

$$\begin{aligned} \mathbf{U}_1 &= -(\omega c/D)(\mathbf{i}\alpha + \mathbf{j}\beta)P, \\ \mathbf{U}_2 &= (i\Omega c/D)(-\mathbf{i}\beta + \mathbf{j}\alpha)P, \\ \mathbf{U}_3 &= i\omega Q\mathbf{k}, \end{aligned} \tag{1}$$

with

$$D = \Omega^2 - \omega^2. \tag{2}$$

The first component is the gradient velocity, the second is the geostrophic; and the third, the thermobaric.

The residual equations are

$$\begin{aligned} c\left(\frac{\mathrm{d}}{\mathrm{d}z} + \Gamma\right)P &= (N^2 - \omega^2)Q, \\ c\left(\frac{\mathrm{d}}{\mathrm{d}z} - \Gamma\right)Q &= (1 + \varkappa^2 c^2/D)P, \end{aligned} \tag{3}$$

and the boundary condition is

$$Q = 0 \quad \text{at} \quad z = 0. \tag{4}$$

The coefficients of Eq. (49–3) are constants when the atmosphere is isothermal, and all stages of the calculation are analogous to those of Chapter VIII, but the algebraic complexity is increased by the presence of the denominator D, rather than ω^2.

Certain conclusions follow at once from an inspection of these equations. For large values of ω/Ω, the denominator D differs from $-\omega^2$ only by terms of the order Ω^2/ω^2 not Ω/ω. Hence the fields P and Q will differ from those of the last chapter only by quantities of this same order, Ω^2/ω^2. Even if the period $2\pi/\omega$ is one hour, the effect of rotation on P and Q will be only about one per cent; for a period of one minute, the effect will be about one thousandth of one per cent.

On the other hand, the ratio of the geostrophic to the gradient velocity is

$$U_2/U_1 = \Omega/\omega$$

and hence the effect of rotation on velocity is much greater: for a one hour oscillation, the velocity components are affected by rotation to the extent of eight per cent. The magnitude, disregarding direction, is affected only to a small degree.

Consequently, for high frequencies, the effect of rotation will be negligible; for intermediate frequencies, its effect on P and Q may be ignored, and its effect will be completely described by Eq. (49–1). Only when ω/Ω is of the order of magnitude unity, or less, will rotation introduce major changes in the eigensolutions.

50. Lamb's waves

The same procedure as before yields Lamb's solutions in the form

$$Q = 0,$$

$$P = \exp(-\Gamma z), \tag{1}$$

$$\omega^2 = \varkappa^2 c^2 + \Omega^2,$$

the graph of the last equation being shown on Fig. 21. The geostrophic component no longer being zero, the hodograph of the waves is a horizontal ellipse. Since Eq. (50–1) shows that $\omega^2 > \Omega^2$ for these waves, its major axis will have the direction of $\mathbf{i}\alpha + \mathbf{j}\beta$, and the ratio of its axes will be Ω/ω. The gradient component always has a greater amplitude than the geostrophic.

Equation (50–1) also shows that the waves are dispersive, having the phase velocity

$$c_{\text{ph}} = \omega/\varkappa = \omega c/(\omega^2 - \Omega^2)^{\frac{1}{2}}. \tag{2}$$

This is always greater than c, and approaches infinity as $\omega \to \Omega$. The group velocity has the direction $\mathbf{i}\alpha + \mathbf{j}\beta$, and its magnitude is

$$c_{\text{gr}} = c^2/c_{\text{ph}}. \tag{3}$$

This is always less than c, and approaches zero as $\omega \to \Omega$.

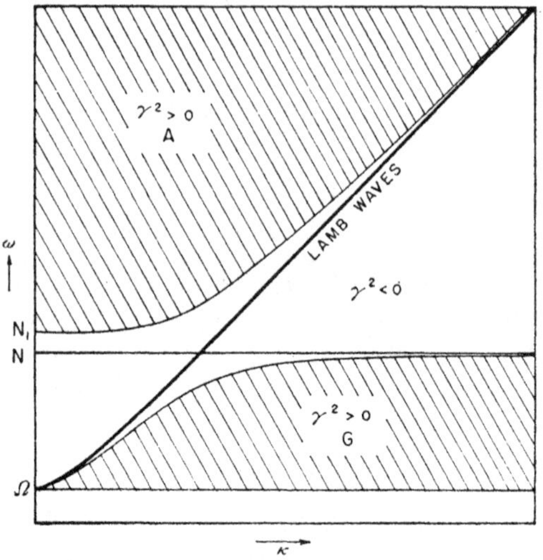

FIG. 21. Diagnostic, or $\varkappa$, ω diagram for an isothermal atmosphere rotating about a vertical axis.

The energy density is

$$E = \omega^2|P|^2/\varkappa^2c^2 \tag{4}$$

and the energy flow is

$$J = \omega|P|^2/\varkappa c = c_{\text{gr}}E \tag{5}$$

in the direction of the horizontal propagation vector. The characteristic impedance is

$$R = \varkappa^2c^2/\omega^2 = (\omega^2 - \Omega^2)/\omega^2 \tag{6}$$

and becomes zero as $\omega \to \Omega$, unity as $\omega \to \infty$.

The dispersion of Lamb's waves is negligible except for frequencies comparable to Ω. Thus, for waves of period one hour, $c_{\text{ph}}/c = 1{\cdot}03$; and for waves of shorter period, the ratio is more nearly unity. The wave-length, $2\pi/\varkappa$, associated to a one-hour wave is about 1200 km. For such long waves, the curvature of the Earth's level surfaces begins to be effective, so that the present calculations are of doubtful applicability to the Earth.

51. Simple waves and eigensolutions

For simple waves, the reduced equations become

$$\begin{aligned} c(i\gamma + \Gamma)P - (N^2 - \omega^2)Q &= 0, \\ (1 + \varkappa^2 c^2/D)P - c(i\gamma - \Gamma)Q &= 0. \end{aligned} \tag{1}$$

In order that these equations have a non-trivial solution, it is necessary that

$$\gamma^2 c^2 = \omega^2 - N_1{}^2 + \varkappa^2 c^2(N^2 - \omega^2)/(\omega^2 - \Omega^2). \tag{2}$$

In order that the solution be regular, γ must again be real, which divides the diagnostic into four areas, as shown on Fig. 21. The area A, of the acoustic waves, is much the same as before; but the area G, of the internal gravity waves, now has the lower boundary $\omega = \Omega$, rather than $\omega = 0$. There is a general correspondence between the eigensolutions with and without rotation. Those on the lower boundary of G in Fig. 21 correspond to some of those on the lower boundary of G in Fig. 16. However, the former have the eigenfrequency $\omega = \Omega$; and the latter, $\omega = 0$. This is closely related to some of the differences between the results of Sections 13 and 15. The solutions obtained in Section 15, for plane level surfaces and vertical Ω, all correspond to the eigenfrequency zero and, therefore, to the $\varkappa$-axis of Fig. 21, not to the lower boundary of area G.

It should be noted that Fig. 21 has been prepared with $N/\Omega = 5$. If $2\pi/\Omega = 12$ hr, this implies that the temperature is more than 100,000°C. A more realistic value would be $N/\Omega = 125$, but a graph prepared with this value would be illegible.

Equation (51–2) is also the equation of the propagation surface. As before, this is an ellipsoid when $\omega^2 > N_1{}^2$ and is an hyperboloid when $\Omega^2 < \omega^2 < N_1{}^2$. In all other frequency ranges, it is imaginary. The phase velocity, for constant ω, is again inversely proportional to the radius vector of the propagation surface, so that there is little to be added to the previous discussion of this matter, except that in the hyperboloidal case, the half-angle of the cone of asymptotes is now

$$\tan\theta = [(\omega^2 - \Omega^2)/(N^2 - \omega^2)]^{\frac{1}{2}} \tag{3}$$

and becomes zero for $\omega = \Omega$.

The general theory of rays and group velocity is the same as before, but the rational Hamiltonian is

$$W = (\alpha^2 + \beta^2)c^2(\omega^2 - N^2) + \gamma^2c^2(\omega^2 - \Omega^2) - (\omega^2 - \Omega^2)(\omega^2 - N_1^2). \qquad (4)$$

The group velocity is again calculable from the equations

$$c_x = -(\partial W/\partial\alpha)/(\partial W/\partial\omega), \qquad \text{etc.}$$

Hence

$$c_x = \frac{\alpha\omega c^2}{\omega^2 - \Omega^2} R,$$

$$c_y = \frac{\beta\omega c^2}{\omega^2 - \Omega^2} R, \qquad (5)$$

$$c_z = \frac{\gamma\omega c^2}{\omega^2 - N^2} R,$$

where

$$R = (\omega^2 - N^2)(\omega^2 - \Omega^2)/[\omega^2(2\omega^2 - N_1^2 - \Omega^2 - k^2c^2)], \qquad (6)$$

is the characteristic impedance. The algebraic identities are

$$\frac{1}{R} = \frac{\omega^2}{\omega^2 - N^2} - \varkappa^2c^2\frac{(N^2 - \Omega^2)\omega^2}{(\omega^2 - \Omega^2)^2(\omega^2 - N^2)} - W\frac{\omega^2}{(\omega^2 - \Omega^2)^2(\omega^2 - N^2)} \qquad (7)$$

$$= \frac{\omega^2}{\omega^2 - \Omega^2} + (\gamma^2 + \Gamma^2)c^2\frac{(N^2 - \Omega^2)\omega^2}{(\omega^2 - \Omega^2)(\omega^2 - N^2)^2} - W\frac{\omega^2}{(\omega^2 - \Omega^2)(\omega^2 - N^2)^2}. \qquad (8)$$

For fixed ω and k, R is again independent of the direction of propagation. For $W = 0$, real γ and $N^2 > \Omega^2$, R is still less than unity. Its dependence on $\varkappa$ and ω is shown on Fig. 22.

The relation between the P and Q fields remains the same as if there were no rotation:

$$Q = [c(i\gamma + \Gamma)/(N^2 - \omega^2)]P. \qquad (9)$$

The pressure–entropy impedance is therefore given by

$$S^2 = (N^2 - \omega^2)^2/[N^2(\gamma^2 + \Gamma^2)c^2]$$

$$= R\,\frac{\omega^2(N^2 - \Omega^2)}{N^2[(1 - R)\omega^2 - \Omega^2]}\,. \tag{10}$$

Its contours differ almost imperceptibly from those of R (Fig. 22) even when $\Omega/N = 5$.

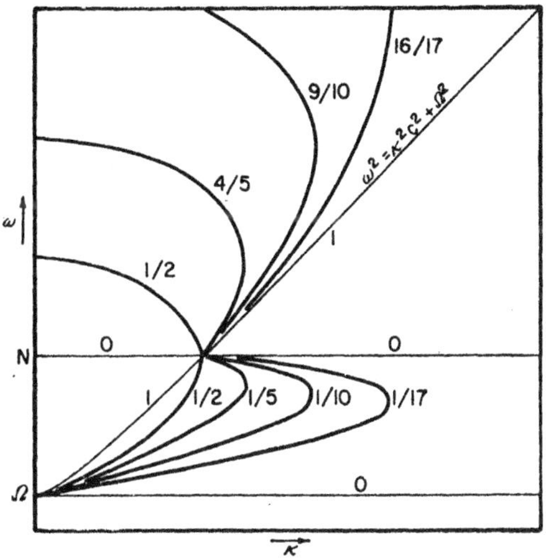

FIG. 22. The dimensionless impedance, R, as a function of $\varkappa$ and ω, for an isothermal atmosphere rotating about a vertical axis.

The components of the velocity are

$$U_x = (\alpha\omega + i\beta\Omega)Pc/(\omega^2 - \Omega^2),$$

$$U_y = (\beta\omega - i\alpha\Omega)Pc/(\omega^2 - \Omega^2), \tag{11}$$

$$U_z = (\gamma - i\Gamma)\omega Pc/(\omega^2 - N^2).$$

The energy density is therefore

$$E = A|P|^2/c \tag{12}$$

with

$$A = \tfrac{1}{2}\left\{1 + \varkappa^2c^2\,\frac{\Omega^2 + \omega^2}{(\omega^2 - \Omega^2)^2} + (\gamma^2 + \Gamma^2)c^2\,\frac{N^2 + \omega^2}{(\omega^2 - N^2)^2}\right\}. \tag{13}$$

The energy flow is

$$J_x = \alpha\omega|P|^2c/(\omega^2 - \Omega^2) = |P|^2c_x/Rc,$$
$$J_y = \beta\omega|P|^2c/(\omega^2 - \Omega^2) = |P|^2c_y/Rc, \tag{14}$$
$$J_z = \gamma\omega|P|^2c/(\omega^2 - N^2) = |P|^2c_z/Rc,$$

so that the rays are again streamlines of the energy flow. The quantities A, R and W are related by the identity

$$2\left(A - \frac{1}{R}\right) = \frac{W}{(\omega^2 - N^2)(\omega^2 - \Omega^2)}\left[\frac{2\omega^2}{\omega^2 - N^2} + \frac{2\omega^2}{\omega^2 - \Omega^2} - 1\right] \tag{15}$$

so that, because $W = 0$,

$$E = |P|^2/Rc, \tag{16}$$

which completes the proof that R is again the characteristic impedance, and that the energy flows with the group velocity.

As in the case of no rotation, the elastic energy is

$$\frac{1}{2c}|P|^2 = \tfrac{1}{2}RE, \tag{17}$$

but the thermobaric energy is

$$\frac{1}{2c}N^2|Q|^2 = \tfrac{1}{2}RE/S^2. \tag{18}$$

Consequently, the kinetic energy is no longer half the total, but

$$\frac{1}{2c}|\mathbf{U}|^2 = \tfrac{1}{2}[1 - R(1 + 1/S^2)]E. \tag{19}$$

As in the case of Lamb's waves, the quantitative differences between these equations and those of Chapter VIII are slight except when ω/Ω is nearly unity. However, reference to Fig. 21 shows that this no longer implies a large wave-length.

The discussion of the eigensolutions and their phase diagrams parallels that of Section 47 and need not be repeated here. The ratio of the geostrophic velocity to the horizontal gradient velocity is Ω/ω, as noted above, and is less than unity for all eigensolutions except those of eigenfrequency zero. The ratio of the vertical thermobaric velocity to the horizontal gradient velocity is

$$|U_3/U_1| = T|\sin \gamma z/\sin (\gamma z - \psi)|$$

where, now

$$T^2 = \frac{(\omega^2 - \Omega^2)(\omega^2 - \varkappa^2 c^2 - \Omega^2)}{\varkappa^2 c^2(\omega^2 - N^2)}. \tag{20}$$

The contours of this function differ from those of Fig. 20 only slightly. The contour $T = 0$ consists of the line $\omega = \Omega$ and the hyperbola $\omega^2 = \varkappa^2 c^2 + \Omega^2$, but the curves for large values of T are essentially unaltered. For small values of $(\omega^2 - \Omega^2)/N^2$, the approximation

$$T^2 = (\omega^2 - \Omega^2)/N^2 \tag{21}$$

is valid, provided $\varkappa c/N$ is not too small. In the interior of area A, the approximation

$$T = \gamma/\varkappa \tag{22}$$

is again valid.

52. Sub-critical stability

In Sections 27, 28 it was noted that $N = \Omega$ is a critical value of the stability. In the last section, it has been supposed that $N/\Omega > 1$. If $N/\Omega < 1$, either the air temperature or the angular velocity must be supposed to have extremely high values. In spite of this, a brief account of the qualitative differences between the two cases is of theoretical interest.

Figure 23 shows the diagnostic diagram for a sub-critical stability. The curve for Lamb's waves remains unaltered, and the shape of area A is not much changed, except that its lower corner is at $\omega = \Omega$, rather than $\omega = N_1$. The shape of area G is quite different, however, than in Fig. 21. The most striking change is in the propagation surfaces for gravity waves when $\Omega > \omega > N$: these are hyperboloids of two sheets (rather than one). The excluded directions of propagation are those outside (rather than inside) the cone of asymptotes. In the normal case, no gravity waves are propagated vertically; in this case, none are propagated horizontally.

These remarks suffice to indicate that the common neglect of stability ($N = 0$) and retention of Coriolis force ($\Omega \neq 0$) is an invalid procedure.

53. Inclined axis of rotation

If the axis of rotation is inclined, all of the preceding calculations become much more complicated, unless the horizontal component of the Coriolis vector $\mathbf{\Omega}$ is neglected (Section 37). If this approximation is made, the calculations of the last four sections need be altered only by replacing Ω with Ω_V in all equations. In this section, some of the simpler effects that depend on Ω_H will be discussed, but no attempt will be made to give a complete account of the solutions of the field equations.

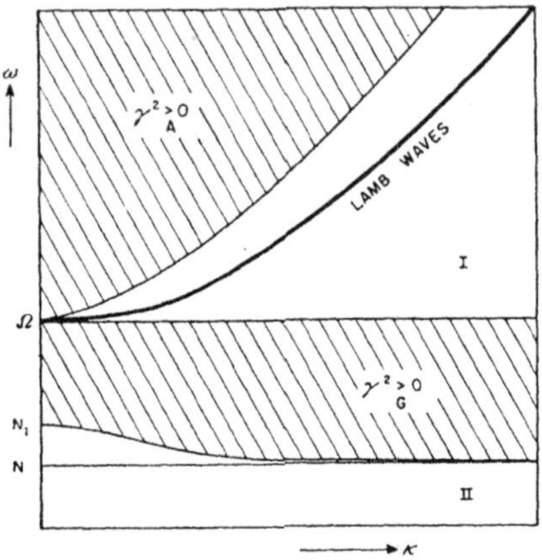

FIG. 23. Diagnostic, or κ, ω diagram, for an isothermal atmosphere with sub-critical stability—$N < \Omega$.

The residual equations, Eqs. (37–18), are

$$c\left(\frac{d}{dz} + \Gamma_1 - i\Gamma_2\right)P = (N^2 - \nu)Q, \tag{1}$$

$$c\left(\frac{d}{dz} - \Gamma_1 - i\Gamma_2\right)Q = (1 - c^2/\mu)P, \tag{2}$$

provided the abbreviations

$$\Gamma_1 = \Gamma - \omega\Omega_H\alpha/(\omega^2 - \Omega_V{}^2), \tag{3}$$

$$\Gamma_2 = \Omega_V\Omega_H\beta/(\omega^2 - \Omega_V{}^2), \tag{4}$$

$$\nu = \omega^2(\omega^2 - \Omega^2)/(\omega^2 - \Omega_V{}^2), \tag{5}$$

$$\mu = (\omega^2 - \Omega_V{}^2)/\kappa^2 \tag{6}$$

are introduced. The boundary condition, $Q = 0$ at $z = 0$, remains unaltered.

One may immediately note that when $\omega \gg \Omega$, these equations differ from Eq. (49–3) by terms that are mostly of the order of magnitude $(\Omega/\omega)^2$. The single exception is that

$$(\Gamma - \Gamma_1)/\Gamma = (\alpha/\Gamma)[\omega\Omega_H/(\omega^2 - \Omega_V{}^2)].$$

This is not only of the order of magnitude Ω/ω, but contains the factor (α/Γ), which may be large. In this frequency range, therefore, the only first-degree effect on P and Q depends on Ω_H, and is neglected when Ω_H is neglected.

The effect of Ω_V on the relation between the velocity $\mathbf{U}$ and the (P,Q) fields has been discussed in Section 49. The additional effect of Ω_H can be seen from Eqs. (37–13, 14): the horizontal pressure gradient, $\nabla_H P$, is to be replaced by

$$\mathbf{G} = \nabla_H P + \mathbf{i}(i\omega\Omega_H Q/c).$$

One may suppose that Eq. (51–9),

$$Q = P/NS$$

will still give an estimate of the ratio of Q to P, so that the ratio of the two terms in $\mathbf{G}$ is of the order of magnitude

$$(\omega/\varkappa c)(\Omega_H/NS).$$

For high-frequency acoustic waves, $(\omega/\varkappa c)$ and S are both of magnitude unity; the additional term thus produces effects of the order of magnitude Ω_H/N. These will be small but still greater than Ω/ω. However, for waves of type G, both $(\omega/\varkappa c)$ and S become small, and no general conclusion is possible.

Lamb's waves. If the atmosphere is isothermal, the coefficients of Eqs. (53–1, 2) are constants, and Lamb's solution becomes

$$Q = 0, \qquad P = \mathrm{e}^{-(\Gamma_1 - i\Gamma_2)z}, \tag{7}$$

$$\mu = c^2, \quad \text{or} \quad \omega^2 = \varkappa^2 c^2 + \Omega_V{}^2. \tag{8}$$

The solution will be regular (finite at $z = \infty$) and an eigensolution only if $\Gamma_1 > 0$: i.e. if

$$\omega\Omega_H\alpha \leqslant \Gamma(\omega^2 - \Omega_V{}^2). \tag{9}$$

In order to discuss Eqs. (53–8, 9) geometrically, it is necessary to generalize the idea of the diagnostic. Instead of the two-dimensional diagram, it is now required to consider a three-dimensional space, in which α, β and ω are the rectangular co-ordinates. It is sufficient to consider only the half-space $\omega > 0$; for the basic equations, Eqs. (37–15, 16), remain unaltered when the four quantities i, α, β and ω are simultaneously replaced by $-i$, $-\alpha$, $-\beta$ and $-\omega$. All descriptions of the three-dimensional diagnostic are therefore to be understood as referring only to its upper half and to be extendable to its

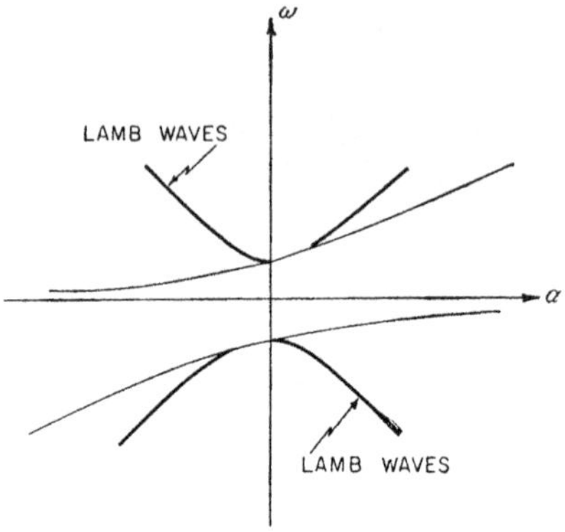

FIG. 24. Illustrating the three-dimensional diagnostic for Lamb's waves in an isothermal atmosphere rotating about an inclined axis.

lower half in accordance with this symmetry principle. It will also be supposed that Ω_H and Ω_V are both positive; then it becomes possible to use geographic terms in a simpler manner than would otherwise be possible.

In this half-space, Eq. (53–8) represents one sheet of an hyperboloid of revolution, and Eq. (53–9) a perforation or hole in this hyperboloid. The perforation is bounded by the intersection of the hyperboloid with the surface $\Gamma_1 = 0$, which is the hyperbolic cylinder

$$\omega\alpha\Omega_H = \Gamma(\omega^2 - \Omega_V^2). \tag{10}$$

When $\Omega_H < \Gamma c$, the perforation is an oval extending eastward from the vortex of the hyperboloid; when $\Omega_H > \Gamma c$, the perforation becomes a slit and extends eastward to infinity. The former case is the normal one and is illustrated by Fig. 24.

As in the case of a vertical axis, Lamb's waves can be considered as unreflected plane waves whose amplitude decreases exponentially with altitude. However, their direction of propagation is not horizontal; their propagation vector is

$$\alpha, \beta, \gamma = -\Gamma_2,$$

and makes an angle δ with the horizontal, where

$$\tan \delta = -\Gamma_2/\varkappa = (\beta/\varkappa)[\Omega_H\Omega_V/(\omega^2 - \Omega_V{}^2)]. \tag{11}$$

For high frequencies, this angle is very small and may be neglected. To this approximation, everything is independent of Ω_H. For east-west azimuths, $\delta = 0$ for all ω, and Ω_H does not enter the formulae so long as the perforation in the hyperboloid may be ignored.

For all other azimuths $\delta \to \pi/2$ as $\omega \to \Omega_V$, and is an essential parameter of the solution. In this limit the Lamb waves are propagated vertically, whereas the neglect of Ω_H would lead one to believe that they are propagated horizontally.

It may appear strange that waves whose propagation vector is inclined to the horizontal boundary are not reflected there. However, these waves are horizontally-polarized, and their energy flows horizontally. It will be seen that related phenomena occur with the other eigensolutions.

The laws of reflection. Equations (53–1, 2) have simple wave solutions; it will be useful to write the vertical component of the propagation vector as

$$\gamma = \gamma_0 + \Gamma_2.$$

Then the residual equations are, in the notation of Section 53

$$\begin{aligned} c(i\gamma_0 + \Gamma_1)P &= (N^2 - \nu)Q, \\ c(i\gamma_0 - \Gamma_1)Q &= (1 - c^2/\mu)P, \end{aligned} \tag{12}$$

and the propagation equation is

$$c(\gamma_0{}^2 + \Gamma_1{}^2) + (N^2 - \nu^2)(1 - c^2/\mu) = 0. \tag{13}$$

For given α, β and ω, this equation has two roots, $\pm\gamma_0$. The boundary condition will be satisfied if

$$Q = \{e^{i(\gamma_0+\Gamma_2)z} - e^{-i(\gamma_0-\Gamma_2)z}\}e^{i(\alpha x+\beta y)}, \tag{14}$$

so that the eigensolutions have this form. The two terms of Eq. (53–14) may still be interpreted as reflected and incident waves, but the angles of incidence and reflection are not equal. The propagation vectors are for the

$$\text{incident wave:} \qquad \mathbf{k}_i = \alpha,\ \beta,\ -(\gamma_0 + \Gamma_2),$$

and for the

$$\text{reflected wave:} \qquad \mathbf{k}_r = \alpha,\ \beta,\ (\gamma_0 - \Gamma_2).$$

They do not have equal lengths, and so there is a change of wave-length on reflection whose amount is determined by Γ_2 and becomes large when $\omega \to \Omega_V$.

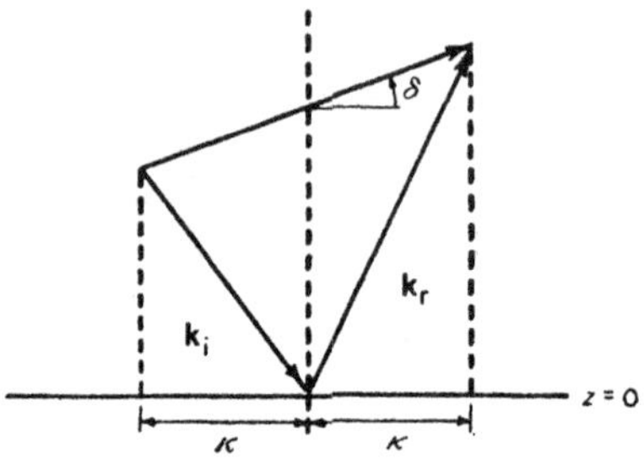

FIG. 25. The reflection of simple waves by a horizontal surface in an isothermal atmosphere rotating about an inclined axis.

This phenomenon is most readily described in terms of the mean propagation vector, $(\mathbf{k}_i + \mathbf{k}_r)/2$. If $\alpha = \varkappa \cos \phi$, $\beta = \varkappa \sin \phi$, this mean has the components

$$\varkappa \cos \phi, \quad \varkappa \sin \phi, \quad \varkappa \tan \delta,$$

where δ is the angle defined by Eq. (53–11). The relations between the incident and reflected propagation vectors are therefore summarized by Fig. 25. When ω/Ω_V is very large, δ is very small; and the change of wave-length, as well as the inequality of the angles of incidence and reflection, may be ignored. As this ratio approaches unity, however, the change ceases to be negligible and depends specifically upon Ω_H.

These anomalous laws of reflection become still more elaborate if the reflecting surface is not level. They are the result of the monoclinic symmetry that results from the inclination of the axis of rotation.

Conclusions. These calculations are by no means complete. It would be possible to discuss the properties of the simple waves—their phase and group velocities, the impedances, etc.—but the algebraic complexities would be great. These two sample calculations indicate one thing, however: there are effects that depend on Ω_H, and these can be very marked for frequencies in the neighbourhood of $\omega = \Omega_V$.

This has been established only for plane level surfaces. If the level surfaces are spheres, Ω_V will depend on latitude. It may be argued that the effects in question will then occur only in a relatively narrow band of latitudes and that this band will be different for each value of ω. At any given latitude, those values of ω for which these effects are large may be relatively unimportant compared to the much larger range of frequencies for which the effects are small.

This is a conjecture, however, and the traditional approximation that neglects the terms in Ω_H is certainly not confirmed by these calculations. This conjecture will be discussed again in Section **64**.

CHAPTER X

OCEANS WITH CONSTANT COEFFICIENTS

54. Introduction

It has been seen that a heavy gas, if its temperature is everywhere the same, will be characterized by constant values of all three of the coefficients N, Γ and c. This is a direct consequence of the perfect gas laws. A liquid, whose thermodynamic properties are quite different from those of a perfect gas, cannot be stratified so that all three of the coefficients are independent of the altitude (or depth). None the less, the mathematical simplicity of constant coefficients is so great that it is of didactic value to study this case, as an introduction to the theory of oceans, even though it is physically impossible.

This chapter will be devoted to the case of an ocean of constant depth D for which N and c have constant values, while $\Gamma = 0$ everywhere. This latter assumption introduces further mathematical simplifications and is justified by the small values of Γ in the Earth's actual oceans. As in the two preceding chapters, it will also be supposed that the level surfaces are plane, and only the case of a vertical axis of rotation will be treated.

Under these conditions, much of the mathematics of Chapter IX applies without change (other than setting $\Gamma = 0$). In particular the theory of simple waves can be taken over in its entirety. Only the construction of eigensolutions out of the simple wave solutions need be altered.

In the isothermal atmosphere, the simple waves were reflected only at its lower, rigid boundary. In the present case, they will be reflected both at the rigid bottom and at the upper free surface. Choosing the co-ordinates so that the undisturbed free surface is at $z = 0$, the bottom will be at $z = -D$. At $z = 0$, the boundary condition is (cf. Eq. (23–9))

$$-i\omega Pc = gW = gi\omega Q, \tag{1}$$

while at $z = -D$, the condition is

$$Q = 0. \tag{2}$$

While the atmospheric theory had only one boundary condition, the oceanic theory has two; and the differences between the two theories all stem from this fact.

Instead of attempting to construct eigensolutions out of simple waves, it is more convenient to start again with the reduced equations in the form

$$c\left(\frac{dP}{dz}\right) = (N^2 - \omega^2)Q, \tag{3}$$

$$c\left(\frac{dQ}{dz}\right) = [1 - \varkappa^2c^2/(\omega^2 - \Omega^2)]P. \tag{4}$$

The eigensolutions must satisfy all four of these equations.

Equation (54–2) will be satisfied if

$$Q = Q_0 \sin \gamma(z + D), \tag{5}$$

and Eq. (54–4) then shows that

$$P = P_0 \cos \gamma(z + D), \tag{6}$$

both P_0 and Q_0 being constants. Equations (54–3, 4) furnish the two equations

$$\begin{aligned} \gamma c P_0 + (N^2 - \omega^2)Q_0 &= 0, \\ [1 - \varkappa^2c^2/(\omega^2 - \Omega^2)]P_0 - \gamma c Q_0 &= 0, \end{aligned} \tag{7}$$

from which the propagation equation

$$\gamma^2c^2(\omega^2 - \Omega^2) + (N^2 - \omega^2)(\omega^2 - \Omega^2 - \varkappa^2c^2) = 0 \tag{8}$$

follows as the condition that P_0 and Q_0 are not both zero.

Thus far, everything is analogous to the atmospheric case, but Eq. (54–1) has not yet been introduced. Combined with Eqs. (54–5, 6) it becomes

$$cP_0 \cos \gamma D + gQ_0 \sin \gamma D = 0,$$

and Eqs. (54–7) reduce this to

$$N^2 - \omega^2 = g\gamma \tan \gamma D. \tag{9}$$

This is a relation between γ and ω that did not enter the atmospheric theory. Solving Eq. (54–8) for γ, and substituting in Eq (54–9), the result is an equation involving only ω and $\varkappa$:

$$N^2 - \omega^2 = \frac{g}{c}\left[\frac{(\omega^2 - N^2)(\omega^2 - \Omega^2 - \varkappa^2 c^2)}{(\omega^2 - \Omega^2)}\right]$$
$$\times \tan\left\{\frac{D}{c}\left[\frac{(\omega^2 - N^2)(\omega^2 - \Omega^2 - \varkappa^2 c^2)}{(\omega^2 - \Omega^2)}\right]\right\}. \qquad (10)$$

No similar relation between ω and $\varkappa$ entered the atmospheric theory. This equation represents a curve (or rather, set of curves) in the diagnostic $(\omega,\varkappa)$ diagram. Only points on these curves correspond to eigensolutions.

The relation between these results and the usual theory of surface waves on a homogeneous, incompressible fluid[1] can be made clear as follows. Since the fluid is to be homogeneous, $N = 0$, and the usual theory also ignores rotation, so that $\Omega = 0$; in this case, Eq. (54–8) reduces to

$$\gamma^2 + \varkappa^2 = \omega^2/c^2. \qquad (11)$$

Since the fluid is incompressible, $c = \infty$, so that this reduces to

$$\gamma^2 = -\varkappa^2$$

or

$$\gamma = \pm i\varkappa. \qquad (12)$$

Setting $N = 0$ in Eq. (54–9), it then becomes

$$-\omega^2 = gi\varkappa \tan(i\varkappa D)$$

or

$$\omega^2 = g\varkappa \tanh(\varkappa D), \qquad (13)$$

which is the ordinary relation between ω and $\varkappa$ for surface waves on shallow water (Ref. 1, Chapter IX).

55. Theory of an homogeneous compressible ocean

Since Eq. (54–10) is quite elaborate, it is useful to consider first the case $N = 0$, $\Omega = 0$, leaving only the velocity of sound, c, as an additional parameter not considered in the customary theory. Since c is to be constant, the rays of the waves will still be straight lines, and they will be reflected according to the usual law of equal angles. Equation (54–11) applies and may be written

$$\omega^2/c^2 = \gamma^2 + \varkappa^2 = k^2 \tag{1}$$

showing at once that $\omega/k =$ const. c. If k is real, the phase velocity is thus independent of frequency, and there is no dispersion.

However, k will not necessarily be real. It will appear that Eq. (54–9),

$$\omega^2 = -g\gamma \tan \gamma D \tag{2}$$

has an infinite number of real solutions for γ, and one imaginary. This imaginary root corresponds to the surface waves. Moreover, it will be seen that these are the analogue of the Lamb waves in the atmosphere.

Before discussing the quantitative aspects of Eqs. (55–1, 2), a graphical survey is essential. This is aided by the introduction of dimensionless quantities defined as follows

$$\begin{aligned} x &= (\gamma D)^2, \\ y &= \omega^2 D/g, \\ y_\varkappa &= \varkappa^2 c^2 D/g, \\ a &= c^2/gD. \end{aligned} \tag{3}$$

The quantity a characterizes the ocean; as is well known, $(gD)^{\frac{1}{2}}$ is the velocity of long surface waves, so that a is determined as the square of ratio of sound velocity to surface-wave velocity. A reasonable value for c is 1·5 km/sec, and for D is 4·5 km. With these values

$$a = 50$$

which is a large number. The following graphs are constructed for a much smaller value in order to illustrate the problem more clearly.

Using Eqs. (55–3), Eq.(55–1) becomes

$$y = ax + y_\varkappa \tag{4}$$

and Eq. (55–2),

$$y = -x^{\frac{1}{2}} \tan (x^{\frac{1}{2}}). \tag{5}$$

For negative values of x, γ and $x^{\frac{1}{2}}$ are both imaginary, but

$$\begin{aligned} f(x) &= -x^{\frac{1}{2}} \tan (x^{\frac{1}{2}}) \\ &= (-x)^{\frac{1}{2}} \tanh [(-x)^{\frac{1}{2}}] \end{aligned} \tag{6}$$

remains real. Consequently, $f(x)$ can be plotted over the whole range $-\infty < x < +\infty$, as has been done on Fig. 26. This graph has an infinite number of branches, shown by the solid curves marked 0, 1, 2, . . . on Fig. 26. These branches have roots at

$$x = (n\pi)^2, \qquad n = 1, 2, 3, \ldots,$$

where n is the ordinal number of the branch; they become asymptotically infinite at

$$x = [(n - \tfrac{1}{2})\pi]^2, \qquad n = 1, 2, 3, \ldots.$$

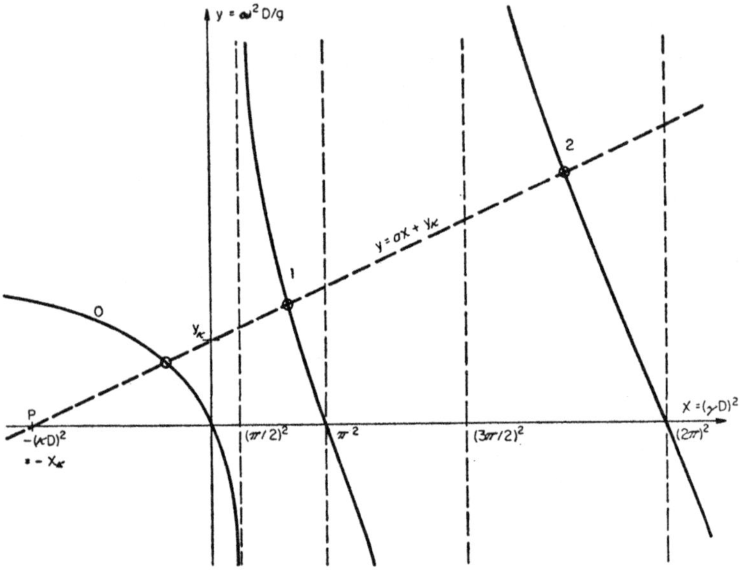

FIG. 26. Diagram for the graphical solution of the simultaneous equations $\gamma^2 + \varkappa^2 = \omega^2/C^2$ and $-\omega^2 = +g\gamma \tan \gamma D$.

For large negative values of x, the branch $n = 0$ is approximated by

$$f(x) = \sqrt{-x} + \ldots, \tag{7}$$

and near the nth asymptote

$$f(x) = \sqrt{x}/[\sqrt{x} - (n - \tfrac{1}{2})\pi] + \ldots. \tag{8}$$

The graph of Eq. (55–4) is shown by the dashed straight line, on Fig. 26, which intersects the x-axis at P, for which

$$x = -(\varkappa D)^2 = -x_{\varkappa}.$$

Its intersections with the various branches of $f(x)$ (indicated by circles) give the values of x and y that satisfy Eqs. (55–4, 5) simultaneously. An inspection of the figures makes it readily possible to trace the variation of these roots as functions of $y_\varkappa$.

Consider first that root which lies on the branch $n = 0$. As $y_\varkappa$ (or $\varkappa$) varies from 0 to $+\infty$, the corresponding value of x (or γ^2) varies from 0 to $-\infty$, while y (or ω^2) varies from 0 to $+\infty$; see the curve $n = 0$ on Fig. 27. As noted above, one is interested primarily

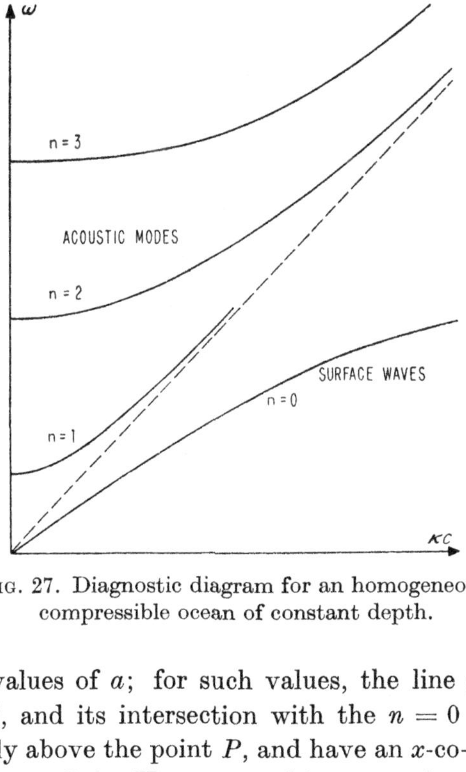

FIG. 27. Diagnostic diagram for an homogeneous, compressible ocean of constant depth.

in very large values of a; for such values, the line $y = ax + y_\varkappa$ is nearly vertical, and its intersection with the $n = 0$ branch will be nearly vertically above the point P, and have an x-co-ordinate nearly equal to $-x_\varkappa = (\varkappa D)^2$. Hence, to this approximation, its y-co-ordinate will be

$$y = \varkappa D \tanh \varkappa D.$$

Since $y = \omega^2 D/g$, the usual formula for surface waves has been recovered (Ref. 1, Chapter IX). It would not be difficult to carry the approximation further and obtain ω as a function of $\varkappa$ expanded in powers of $1/a$.

Consider next the root that lies on the nth branch of $f(x)$. When $y_\varkappa = 0$, its x-co-ordinate is the nth root of the equation

$$a\sqrt{x} = -\tan\sqrt{x}. \tag{9}$$

As $y_\varkappa$ increases, this x-co-ordinate diminishes and approaches $(n - \frac{1}{2})^2\pi^2$ for $y_\varkappa = \infty$; simultaneously, the corresponding value of y increases to infinity; see the curves $n = 1, 2, \ldots$ on Fig. 27. Again, one is primarily interested in large values of a: for these, the roots of Eq. (55–9) differ very little from the asymptotic value

$$x = (n - \tfrac{1}{2})^2\pi^2, \qquad n = 1, 2 \ldots$$

or

$$\gamma D = (n - \tfrac{1}{2})\pi. \tag{10}$$

The z-component, γ, of the propagation vector is thus almost independent of $\varkappa$, and Eq. (55–10) is the so-called "open organ pipe" formula. The corresponding value of y is obtained from Eq. (55–4) and is

$$y = (n - \tfrac{1}{2})^2\pi^2 a + y_\varkappa$$

and does depend on $y_\varkappa$; this equation may be rewritten

$$\omega^2/c^2 = \{(n - \tfrac{1}{2})^2\pi^2/D^2 + \varkappa^2\}, \tag{11}$$

and is well known in the theory of underwater sound.

Again, it is not difficult to carry these calculations to higher orders of $1/a$, and thus to investigate the effect of gravity on the propagation of sound in an ocean of these hypothetical properties. It is interesting to note that these effects are of the order of magnitude

$$gD/c^2$$

which is small. They do not dominate over the purely elastic effects until $D > c^2/g$; since $c^2/g = 225$ km, all of the Earth's oceans are "shallow" in this respect.

Before proceeding, it is important to note the significance of the ordinal number, n, that has appeared in the above discussion. Since

$$Q = Q_0 \sin\gamma(z + D),$$
$$P = P_0 \cos\gamma(z + D),$$

and γ is real for $n > 0$, it follows that both P and Q will have roots at certain values of z. The number of these roots, or nodal surfaces, as z varies from $-D$ to 0, is exactly n: consequently n is sometimes called the nodal number, though the term ordinal number will be used in this book. This interpretation of the ordinal number extends to the case $n = 0$, for then

$$Q = iQ_0 \sinh (i\gamma)(z + D),$$

$$P = P_0 \cosh (i\gamma)(z + D),$$

and P has no root although Q does have one root at $z = -D$.

The nodal surfaces also enter into the description of the fluid motion. On those planes where $Q = 0$, the velocity is horizontal; on those where $P = 0$, the motion is vertical. Moreover, since the energy flux is $\mathbf{J} = P\mathbf{U}$, there is no flow of energy across the planes $P = 0$.

It follows that all the eigensolutions for a given n represent modes of motion that are qualitatively similar; and even though they may have different values of $\varkappa$, γ and ω, they coincide in the number of nodal planes. It is convenient and customary to group together all eigensolutions having the same n, and to call them a "mode"—more properly, a mode of motion or of vibration.

56. Theory of a stratified but incompressible ocean

Because of the shallowness of the Earth's oceans, it is of importance to treat the case $a = \infty$ next. This amounts to setting $c = \infty$ in the equations of Section 54, and treating the ocean as an incompressible fluid. The discussions of this section are very closely related to Love's theory of wave motion in a heterogeneous liquid;[(2)] see also Ref. 1, p. 378).

In this limit, Eq. (54–8) becomes

$$\gamma^2(\omega^2 - \Omega^2) - \varkappa^2(N^2 - \omega^2) = 0, \tag{1}$$

and Eq. (54–9),

$$\omega^2 = N^2 - g\gamma \tan \gamma D. \tag{2}$$

Again, dimensionless quantities, defined by

$$x = (\gamma D)^2, \qquad x_\varkappa = (\varkappa D)^2 \tag{3}$$

$$y = \omega^2 D/g, \qquad y_N = N^2 D/g, \qquad y_\Omega = \Omega^2 D/g,$$

simplify the analysis. Equations (56–1) and (56–2) may then be written

$$x = x_\varkappa(y_N - y)/(y - y_\Omega), \tag{4}$$

$$y = y_N - \sqrt{x} \tan \sqrt{x}. \tag{5}$$

The graph of Eq. (56–5) is entirely similar to that of Eq. (55–5) and consists of the branches $n = 0, 1, 2 \ldots$ which are shown on Fig. 28; however, the curves are now shifted with respect to the x-axis

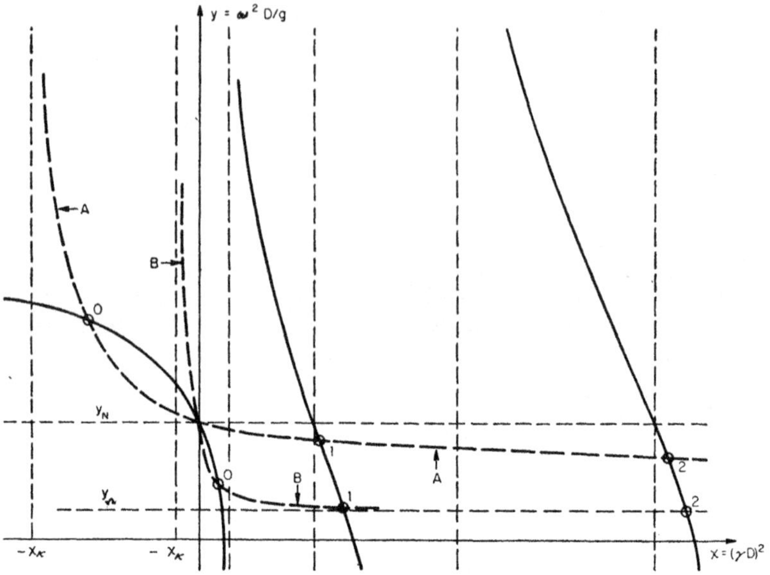

FIG. 28. Diagram for the graphical solution of the simultaneous equations

$$\gamma^2(\omega^2 - \Omega^2) = \varkappa^2(N^2 - \omega^2),$$
$$(N^2 - \omega^2) = g\gamma \tan \gamma D.$$

because $y_N \neq 0$. The graph of Eq. (56–4) is an hyperbola, and two cases have to be distinguished:

case A: $$x_\varkappa > y_N - y_\Omega,$$

and case B: $$x_\varkappa < y_N - y_\Omega.$$

An example of each of these cases is shown on Fig. 28. In case A, the intersection of the hyperbola with the branch $n = 0$ occurs for negative values of x; in case B, this intersection has $x > 0$.

This criterion for case A or B may be rewritten in dimensioned form as

$$\varkappa^2 \gtrless (N^2 - \Omega^2)/gD. \tag{6}$$

Here Ω^2 may be neglected in comparison to N^2, and it becomes apparent that the critical wave number is $N/\sqrt{gD}$. This is the "deep water" wave number associated to the frequency N. To obtain an estimate of its order of magnitude, one may take $D = 4$ km, so that

$\sqrt{gD} = 0{\cdot}2$ km/sec; and suppose that $N = 1$ hr—a value that corresponds to the stability below the permanent thermocline of the Earth's oceans. The critical wave-length is then 720 km, case A occurring for wave-lengths less than this.

The distinction between cases A and B is of importance only for surface waves which correspond to the intersection of the hyperbola with the branch $n = 0$ of Eq. (56–5). When $x_\varkappa$ is very large, it is seen that $y \gg y_N$; consequently Eq. (56–4) yields the approximation $x = -x_\varkappa = -\varkappa^2 D^2$, and Eq. (56–5) then turns into the classic formula

$$\omega^2 = g\varkappa \tanh \varkappa D.$$

To this approximation, the stability has no effect on the surface waves.

However, as $\varkappa$ diminishes and approaches its critical value, there are departures from the classic formulae; and when $\varkappa$ reaches its critical value, x becomes zero. For the larger values of $\varkappa$, Q and P both reach their largest values at the free surface of the fluid, but for the critical value, $Q = 0$ and $P =$ constant for all values of z. Also, for case A, $\omega \geqslant N$, reaching the value N for the critical value. For values of $\varkappa$ less than the critical, $\omega < N$, and x is positive. Therefore γ is real, and Eq. (54–6) shows that P assumes its largest value at the bottom of the fluid; Q still reaches its maximum at the free surface. Finally, for $\varkappa = 0$, it is seen that $\omega = \Omega$, while γ is determined by the equation

$$(N^2 - \Omega^2) = g\gamma \tan \gamma D.$$

In fact, when $\omega < N$, the solutions for $n = 0$ are internal gravity waves rather than surface waves in the classical sense. The relation between ω and $\varkappa$ thus deduced is summarized graphically as the branch $n = 0$ on Fig. 29.

The other roots of Eqs. (56–4, 5) for $n > 0$ are not subject to similar qualitative changes as $\varkappa$ passes through its critical value. It is seen from Fig. 28 that, as $x_\varkappa$ increases from 0 to ∞, y increases from y_Ω to y_N, or, ω from Ω to N. Over this whole range, the solutions retain the character of internal gravity waves. Again, Fig. 29 summarizes the relation between ω, $\varkappa$ and n; each of the curves on this figure is thus associated to a single mode.

In closing this section, it may be well to call attention to another detail of Fig. 28. The hyperbolae all intersect the branch $n = 0$ at

$x = 0$, $y = y_N$, regardless of the value of $x_\varkappa$. This point represents a solution of Eqs. (56–4, 5) but leads only to a trivial solution of the basic equations, Eqs. (56–3, 4); for Eq. (56–5) shows that for $\gamma = 0$, $Q = 0$ for all z, and Eq. (56–7) shows that $P = 0$.

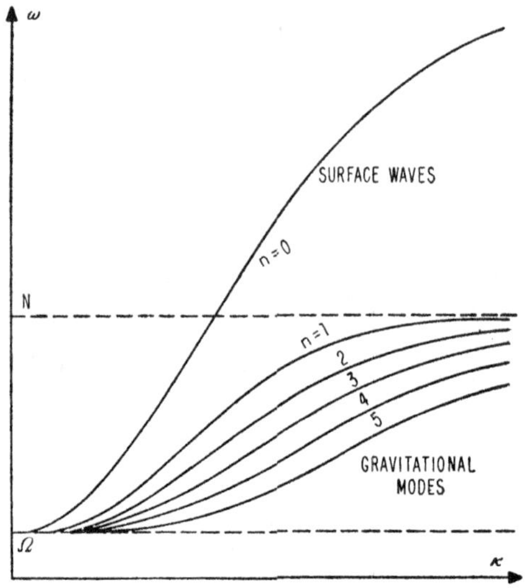

FIG. 29. Diagnostic diagram for a stratified but incompressible ocean of constant depth.

57. The general case

Having discussed the special cases of the two preceding sections in some detail, the general problem formulated in Section 54 can be treated with comparative brevity. In terms of the dimensionless quantities already defined, Eq. (54–8) becomes

$$ax(y - y_\Omega) + (y_N - y)(y - y_\Omega - y_\varkappa) = 0, \tag{1}$$

and Eq. (54–9) is again

$$y = y_N - \sqrt{x} \tan \sqrt{x}. \tag{2}$$

Equation (57–1) represents an hyperbola; but whereas, in the last section, the corresponding equation led to a rectangular hyperbola, the present case has oblique asymptotes. These are the horizontal line

$$y = y_\Omega, \tag{3}$$

and the inclined line

$$y = ax + y_\varkappa + y_N. \tag{4}$$

This last equation should be compared with Eq. (55–4), to which it reduces when $y_N = 0$.

As shown on Fig. 30, this hyperbola intersects the y-axis at

$$y = y_\varkappa + y_\Omega$$

and two cases present themselves that are similar to those encountered in Section 56. The criterion is now

$$\begin{aligned} &\text{case A:} \qquad y_\varkappa + y_\Omega > y_N, \\ &\text{case B:} \qquad y_\varkappa + y_\Omega < y_N. \end{aligned} \tag{5}$$

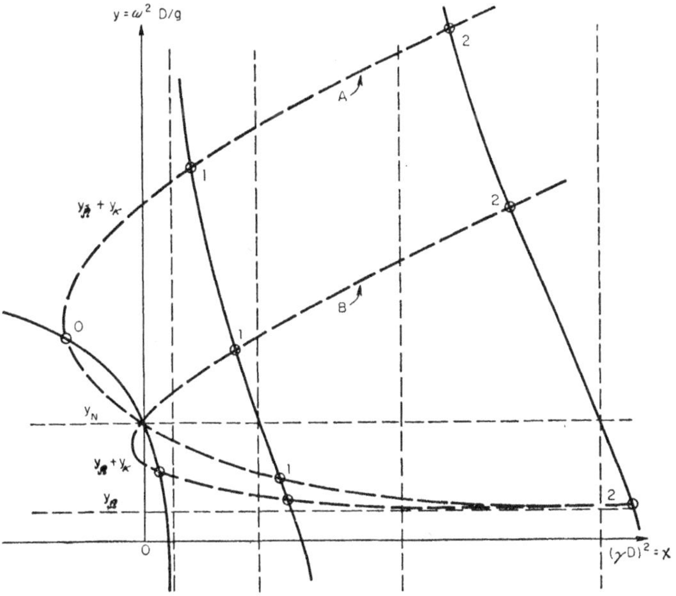

FIG. 30. Diagram for the simultaneous solution of the simultaneous equations

$$\gamma^2 c^2(\omega^2 - \Omega^2) = (N^2 - \omega^2)(\varkappa^2 c^2 - \omega^2 + \Omega^2),$$
$$(N^2 - \omega^2) = g\gamma \tan \gamma D.$$

As above, in case A, the intersection with the branch $n = 0$ occurs for $x < 0$, while in case B, $x > 0$. The intersection at $x = 0$, $y = y_N$ again leads only to the trivial solution of the basic equations.

Inspection of Fig. 30 shows that there are now two roots of Eqs. (57–1, 2) for each value of $n > 0$. One of these corresponds to the intersection of the upper branch of the hyperbola with the nth branch of Eq. (57–2), the other to the lower branch of the hyperbola. The former is an acoustic wave; the latter, a gravity wave. For systematic reasons, that will become clearer below, positive values of n

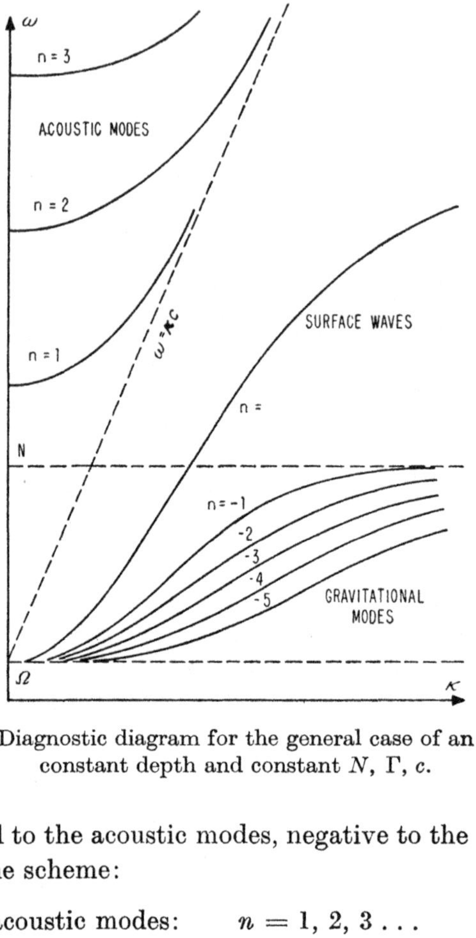

FIG. 31. Diagnostic diagram for the general case of an ocean of constant depth and constant N, Γ, c.

will be assigned to the acoustic modes, negative to the gravity modes, according to the scheme:

acoustic modes:	$n = 1, 2, 3 \ldots$
surface mode:	$n = 0$
gravity modes:	$n = -1, -2, -3 \ldots$

In this way, there is again a one-to-one correspondence between the ordinal numbers and the modes.

This relation between n, ω and $\varkappa$ is shown on Fig. 31 in a qualitative manner; for the sake of clarity, the spacing of the acoustic modes has been much reduced and the slopes of the other curves much increased. Inspection of this figure shows a general correspondence with the diagram for the isothermal atmosphere, Fig. 21. The acoustic modes are in the area A of Fig. 21, and the gravitational modes are in the area G. After noting this, it will be clear that the surface waves of the oceanic case correspond to the Lamb waves of the atmosphere. This correspondence is not confined to the general location of the $(\omega,\varkappa)$ curve with respect to the other modes; both kinds of waves are characterized by a maximum amplitude at a horizontal boundary and by the absence of any horizontal nodal surface. Both are intimately connected with the existence of interfaces and may be called interfacial waves. Similar interfacial waves occur in the theory of elastic waves in solids; these are known as Love and Rayleigh waves.

58. A simple approximation for the internal gravity modes

A considerably simpler approximation can be obtained from an inspection of Fig. 30. This shows that for $\omega < N$ and $n < 0$,

$$\gamma D = (|n| + \Delta)\pi \tag{1}$$

where Δ is a function of both ω and n. It is also seen that $\Delta \to 0$ when $\omega \to N$ or when $|n| \to \infty$. Consequently, over a considerable part of the $(\varkappa,\omega)$ diagram below $\omega = N$, a reasonable approximation is $\Delta = 0$. For more precise results, methods for the calculation of Δ can be devised. Borrowing a term from the theory of atomic spectra, Δ may be called the ordinal defect; this is a concept that will recur in the following sections.

Solving Eq. (54–8) for γ and using Eq. (58–1),

$$\{(N^2 - \omega^2)[\varkappa^2 c^2/(\omega^2 - \Omega^2) - 1]\}^{\frac{1}{2}} = (|n| + \Delta)\pi c/D, \tag{2}$$

which is a rigorous relation between ω, $\varkappa$ and n, provided the ordinal defect is calculated. As noted, reasonable approximation is often obtained by setting $\Delta = 0$. If the additional approximation $c = \infty$ is used, the simple formula

$$\varkappa\left(\frac{N^2 - \omega^2}{\omega^2 - \Omega^2}\right)^{\frac{1}{2}} = |n|\pi/D \tag{3}$$

is obtained.

Figure 32 shows graphs of this equation for selected values of $n < 0$.

and $D = 4$ km, $2\pi/N = 15$ min, $2\pi/\Omega = 12$ hr. The graph of Eq. (54–3) for $n = 0$ has been included for comparison.

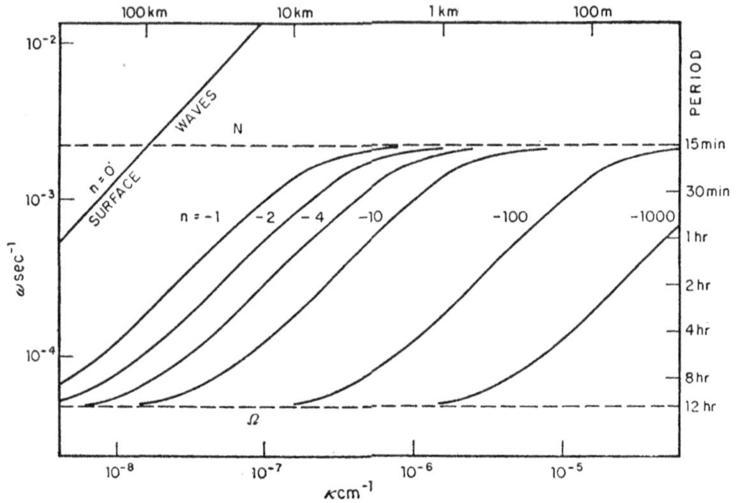

FIG. 32. Portion of the diagnostic diagram for $D = 4$ km, $2\pi/N = 15$ min, $2\pi/\Omega = 12$ hr.

59. The modes of a rectangular tank

In Chapter VI, it was noted that the eigensolutions for a fluid of finite volume usually are quite different from those of an infinite volume. In both cases the energy density is everywhere finite, but in the latter case, the total energy is often infinite. This reflects itself in the number of eigensolutions: for a finite volume, their number is a denumerable infinity; for an infinite volume, it is often a non-denumerable infinity.

In Section 35, the decision was made to treat only fluids bounded by level surfaces. The absence of lateral boundaries implies an infinite volume. In this section, the effects of the lateral boundaries will be examined, though not exhaustively.

Consider a rectangular tank, with level bottom, free surface and vertical walls. Let its horizontal dimensions be A and B, its depth D; the co-ordinate axes may then be chosen in the edges of the tank. In addition to the boundary conditions

$$Pc + gQ = 0 \quad \text{at} \quad z = 0, \tag{1}$$

$$Q = 0 \quad \text{at} \quad z = D, \tag{2}$$

the conditions at the vertical walls are

$$U = 0 \quad \text{at} \quad x = 0 \text{ and } x = A, \tag{3}$$

$$V = 0 \quad \text{at} \quad y = 0 \text{ and } y = B. \tag{4}$$

In order to avoid the anomalous reflections treated in Section 53, it will be supposed that $\Omega = 0$. The eigensolutions can then be constructed out of plain waves that are reflected at the walls. They will have the form

$$\begin{aligned} P &= P_0 \cos \alpha x \cos \beta y \cos \gamma(z + D), \\ Q &= Q_0 \cos \alpha x \cos \beta y \sin \gamma(z + D), \\ U &= -\alpha c P_0 \sin \alpha x \cos \beta y \cos \gamma(z + D), \\ V &= -\beta c P_0 \cos \alpha x \sin \beta y \cos \gamma(z + D). \end{aligned} \tag{5}$$

Equations (59–3) and (59–4) will be satisfied if

$$\alpha = k\pi/A, \qquad \beta = m\pi/B, \tag{6}$$

where k and m are positive integers. Correspondingly,

$$\varkappa = (\alpha^2 + \beta^2)^{\frac{1}{2}} = \pi[(k/A)^2 + (m/B)^2]^{\frac{1}{2}}. \tag{7}$$

The effect of the vertical walls is immediately apparent: without walls, α, β and $\varkappa$ can vary continuously. With walls, they are restricted to a series of discrete values, determined by the integers k and m.

Once $\varkappa$ has been fixed, however, the calculations given in the earlier sections of this chapter proceed unaltered. The ordinal number n is still given by Eq. (58–1):

$$\gamma = (|n| + \Delta)\pi/D, \tag{8}$$

whose resemblance to Eq. (59–6) will be apparent. The integers k and m may also be called ordinal numbers: they determine the number of vertical nodal planes of P and Q in much the same way that n determines their horizontal nodes.

These matters find their representation in the diagnostic diagram. For simplicity, suppose the fluid is incompressible, so that the calculations of Section 56, with $\Omega = 0$, apply. The nodal curves are then shown by light lines on Fig. (33). However, not all values of $\varkappa$ are consistent with Eq. (59–6); those that are consistent are shown by the vertical lines. Hence eigensolutions correspond only to those values of $\varkappa$ and ω determined by the intersections of the curves with the vertical lines.

The discrete values of the frequency, ω, that are so determined are variously known as resonant frequencies, natural frequencies or free frequencies.

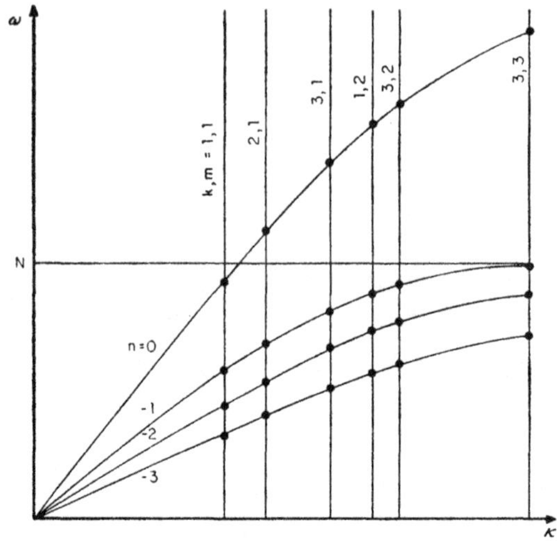

FIG. 33. Diagnostic diagram for a rectangular tank.

The spacing of the vertical lines on Fig. 33 has been fixed for the example $A = 2B$. Only those for k, m = 1, 2, 3 have been drawn, but there are an infinite number of them, and their spacing tends to become closer as k and m increase. Correspondingly, the spacing of the resonant frequencies becomes closer as the ordinal numbers increase. The frequency $\omega = N$ is a limit point of the set of resonant frequencies. Again borrowing a term from quantum theory, those resonant frequencies that have the same value of the vertical ordinal n, but different values of the horizontal ordinals k, m, may be said to form a series.

60. Other lateral boundaries

The mathematical problems that arise when the fluid has non-level boundaries of various shapes are numerous, and usually difficult. The case of free steady flow in a basin of arbitrary bottom topography was discussed in Section 13, and it was seen that there was a boundary layer, in which the flow tended to be bathystrophic. Outside this layer, the influence of the boundary persists, but is not dominant.

These complications have been avoided either by assuming that all boundaries are level, or, as in the last section, that the lateral boundaries are vertical.

In general, the problems raised by vertical lateral boundaries, even if their shape is arbitrary, are less difficult than those raised by basins of more natural shapes. In particular the residual equations Eq. (36–12) yield the dependence of the eigensolutions on the vertical. co-ordinates, and this dependence is the same at all places in the tank Their dependence on the horizontal co-ordinates is governed by the two-dimensional wave equation, (Eq. 36–11), which is

$$\frac{\partial^2 P_1}{\partial x^2} + \frac{\partial^2 P_1}{\partial y^2} + \varkappa^2 P_1 = 0. \tag{1}$$

This must be solved subject to the conditions that the normal component of the velocity vanishes at the vertical walls. For a rectangular tank, this solution was obtained in the preceding section. The solution is known for tanks of circular, elliptic and parabolic sections (Ref. 1, Ch. VIII). For other shapes, the problem can be approached numerically.

References for Chapter X

1. H. LAMB, *Hydrodynamics*, Cambridge Univ. Press, 6th ed., 728 pp., 1932.
2. A. E. H. LOVE, Wave motion in a heterogeneous heavy liquid, *Proc. Lond. Math. Soc.* **22,** 307 (1891).

CHAPTER XI

GENERAL THEORY OF RAYS

61. Introduction

THE further development of the theory requires a decision at this point. One may either proceed with the theory of simple waves and the rays associated with them or with the theory of the eigensolutions, based on the residual equations. This chapter will be devoted largely to the theory of rays, and the general discussion of the residual equations will not be resumed until Chapter XIV. The intervening chapters will be devoted largely to illustrations of the ray theory, although they will also provide occasion for deriving some special results from the residual equations.

There are two kinds of ray theory whose results are sometimes identical, sometimes very different. One is based on the theory of the characteristic curves and surfaces associated with partial differential equations. Its results are always rigorous and of importance for the theory of shock waves and for certain technical purposes (Ref. 10, p. 526; 3, p. 342; 1, p. 132).

The other theory is rigorous only when the differential equations have constant coefficients. This is the theory that has already been applied to the isothermal atmosphere in the previous chapters. In order to extend it to the case of variable coefficients, it is necessary to use a method of approximation usually associated with the names of Wentzel,[12] Kramers[8] and Brillouin[5] although a full account of its history would include references to Huyghens, Hamilton, Kirchhoff, de Broglie and many others. This W-K-B method is rigorous only when the coefficients are constants, as has been said, but it yields surprisingly good results in more general cases.[6],[7] This approximation, when applied to the simple wave equation

$$c\,\nabla^2\psi - \partial^2\psi/\partial t^2 = 0, \tag{1}$$

leads to Huyghens' historic construction for wave fronts, which was the first formulation of wave theories. The results to be obtained from its application to the field equations are therefore generalizations of Huyghens' principle even though their geometric interpretation is much more elaborate.

In recent years there has been a revived interest in the W-K-B approximations, and noteworthy advances have been made. Especially interesting from the purely mathematical standpoint is the work of Bremmer[4] and Bellman and Kalaba.[2]

Since the two ray theories lead to different results when applied to the present field equations, it would be desirable to examine both. Space considerations preclude this, and only the second will be developed here. It may be conjectured that the first would lead to a mathematical theory of frontal disturbances in the atmosphere.

62. The Hamilton-Jacobi equation

When the coefficients of the field equations are constant, it has been seen that all of the simple wave fields have the form

$$P = P_0 \exp(i\Phi), \text{ etc.} \tag{1}$$

where

$$\Phi = \alpha x + \beta y + \gamma z - \omega t, \tag{2}$$

and P_0, α, . . . ω, are real constants. The function Φ is the phase and P_0 . . . , are the amplitudes of the wave. It is an obvious conjecture that when the coefficients are variable, the solutions may have this same form but with Φ a real, non-linear function of x, y, z, t and P_0 . . . also variable. In fact, the first equation imposes no restriction on P and the proposition must be made more precise before it ceases to be trivial.

If the first equation is differentiated, the result is

$$\frac{\partial P}{\partial x} = iP\frac{\partial \Phi}{\partial x} + \frac{\partial P_0}{\partial x}\exp(i\Phi),$$
$$\cdot \qquad \cdot \qquad \cdot$$
$$\cdot \qquad \cdot \qquad \cdot$$
$$\frac{\partial P}{\partial t} = iP\frac{\partial \Phi}{\partial t} + \frac{\partial P_0}{\partial t}\exp(i\Phi).$$

It may be that the second term on the right of each of these equations is negligible compared to the first—this is the stronger form of the conjecture and is the W-K-B approximation.

The notation

$$\alpha = \frac{\partial \Phi}{\partial x}, \quad \beta = \frac{\partial \Phi}{\partial y}, \quad \gamma = \frac{\partial \Phi}{\partial z}, \quad \omega = -\frac{\partial \Phi}{\partial t}, \quad \mathbf{k} = \nabla \Phi. \tag{3}$$

will be used. When Φ is a linear function of x, y, z, t, (Eq. 62–2), this notation is identical with that already used for simple waves, and $\mathbf{k}$ is the wave number or propagation vector while ω is the fiequency. When Φ is a more general function, $\alpha \ldots \omega$ are no longer constants but are functions of x, y, z, t; then $\mathbf{k}$ is the local propagation vector, ω, the local frequency. The W-K-B approximation then leads to such equations as

$$\nabla P = i\mathbf{k}P, \qquad \partial P/\partial t = i\omega P$$
$$\nabla \cdot \mathbf{U} = i\mathbf{k} \cdot \mathbf{U}, \quad \text{etc.}$$

On introducing these approximations into the field equations, they become

$$\begin{aligned} i\omega\mathbf{U} - \boldsymbol{\Omega} \times \mathbf{U} &= cP(i\mathbf{k} + \Gamma\boldsymbol{\zeta}) - N^2Q\boldsymbol{\zeta}, \\ c(i\mathbf{k} - \Gamma\boldsymbol{\zeta}) \cdot \mathbf{U} &= i\omega P, \\ \boldsymbol{\zeta} \cdot \mathbf{U} &= i\omega Q. \end{aligned} \tag{4}$$

If the phase function is known, these are linear algebraic equations in $\mathbf{U}$, P, Q. Since they are homogeneous, they will have non-trivial solutions only if their determinant vanishes (cf. Eqs. (42–1, 2). This determinant will depend on $\alpha \ldots \omega$, and the condition that it vanishes will therefore result in a differential equation for the phase Φ, because of Eq. (62–3). This is known as the Hamilton–Jacobi differential equation associated to the field equations.

63. Plane level surfaces: vertical axis

In the first instance this equation will be derived only for the case of plane level surfaces and a vertical axis of rotation. Then Eqs. (61–4), written in component form, are

$$\begin{aligned} i\omega U + \Omega V &= i\alpha cP, \\ i\omega V - \Omega U &= i\beta cP, \\ i\omega W &= (i\gamma + \Gamma)cP - N^2Q, \\ i\omega P &= c[i\alpha U + i\beta V + (i\gamma - \Gamma)W], \\ i\omega Q &= W. \end{aligned} \tag{1}$$

Straight-forward algebraic manipulation reduces these to

$$\begin{aligned} U &= -c(\alpha\omega + i\beta\Omega)P/(\Omega^2 - \omega^2), \\ V &= -c(\beta\omega - i\alpha\Omega)P/(\Omega^2 - \omega^2), \\ W &= i\omega Q; \end{aligned} \tag{2}$$

$$\begin{gathered} [i\gamma + \Gamma]P - (N^2 - \omega^2)Q = 0, \\ [1 - c^2(\alpha^2 + \beta^2)/(\omega^2 - \Omega^2)]P + [i\gamma - \Gamma]Q = 0. \end{gathered} \tag{3}$$

Equations (63–2) clearly are approximations to the earlier expressions for the velocity in terms of the fields P and Q. Equations (63–3) are, similarly, approximations to the residual equations. In order that P and Q shall not both vanish, the determinant of Eq. (63–3) must vanish. This leads to a propagation equation that is formally identical with Eq. (51–2):

$$\gamma^2c^2 = (\omega^2 - N_1^{\,2}) + c^2(\alpha^2 + \beta^2)(N^2 - \omega^2)/(\omega^2 - \Omega^2). \tag{4}$$

Clearing this of fractions and substituting from Eq. (61–3), it is seen to be a partial differential equation for the phase:

$$\begin{aligned} &c^2\left[\left(\frac{\partial\Phi}{\partial x}\right)^2 + \left(\frac{\partial\Phi}{\partial y}\right)^2\right]\left[\left(\frac{\partial\Phi}{\partial t}\right)^2 - N^2\right] \\ &\qquad + \left[N_1^{\,2} - \left(\frac{\partial\Phi}{\partial t}\right)^2 + c^2\left(\frac{\partial\Phi}{\partial z}\right)^2\right]\left[\left(\frac{\partial\Phi}{\partial t}\right)^2 - \Omega^2\right] = 0. \end{aligned} \tag{5}$$

This is the Hamilton–Jacobi equation associated to the field equations.

General and complete solutions. Being a partial differential equation, the most general solution of the Hamilton–Jacobi equation will involve arbitrary functions. For example, the values of Φ at $t = 0$ can be specified at all points of space, and Eq. (63–5) will then uniquely determine it at all other times.

However, less general solutions than these will suffice for the present purposes. If a special solution of Eq. (63–5) contains four arbitrary constants (four being the number of independent variables), it is called a complete solution. It will be sufficient to find a complete solution of this equation; there are geometric methods for constructing the general solution once a complete solution has been obtained, but it will not be necessary to discuss these.

Since Eq. (63–5) involves only the partial derivatives of Φ, it follows that if Φ is one solution, so will be Φ + constant. One of the four constants in the complete solution is therefore an additive constant and need not be considered further.

The Hamilton–Jacobi function. In the present case, the determination of the complete solution is rendered easy by the fact that c, N and N_1 are independent of x, y, t and depend only on z. Because of this, Eq. (63–5) will have solutions of the form

$$\Phi = \alpha x + \beta y - \omega t + S(z), \tag{6}$$

where α, β, ω are now again constants, and the function S satisfies the ordinary differential equation

$$\varkappa^2 c^2(\omega^2 - N^2) + (\omega^2 - \Omega^2)\left[N_1{}^2 - \omega^2 + c^2\left(\frac{dS}{dz}\right)^2\right] = 0, \tag{7}$$

$$(\varkappa^2 = \alpha^2 + \beta^2, \text{ as always})$$

which results from the substitution of Eq. (63–6) into (63–5). Since Eq. (63–6) contains three arbitrary, non-additive constants, a complete solution has thus been defined.

Solving Eq. (63–7) for (dS/dz), we obtain

$$\frac{dS}{dz} = \left\{\varkappa^2 \frac{N^2 - \omega^2}{\omega^2 - \Omega^2} - \frac{N_1{}^2 - \omega^2}{c^2}\right\}^{\frac{1}{2}} = \gamma(\alpha, \beta, \omega, z), \tag{8}$$

which can be integrated at once:

$$S = \int_{z_1}^{z} \gamma \, dz. \tag{9}$$

In this equation the lower limit, z_1, essentially determines the fourth additive constant in the complete solution and may be chosen in any one of a number of ways. However, since S is to be a real function, it is necessary to choose z_1, so that γ is real over the whole range from z_1 to z.

64. The rays and group velocity

Derivation of the rays. The equation of the rays can be derived from the Hamilton–Jacobi function Φ by much the same method that was used in Section 44. In real notation, the fields of a simple wave are

$$P = \sin [\alpha x + \beta y - \omega t + S(\alpha, \beta, \omega, z)], \text{ etc.} \tag{1}$$

Consider the superposition of two such waves, both having the same amplitude but with the parametric values α', β', ω' and α'', β'', ω'':

$$\begin{aligned} P &= \sin [\alpha' x + \beta' y - \omega' t + S'] \\ &+ \sin [\alpha'' x + \beta'' y - \omega'' t + S'']. \end{aligned} \tag{2}$$

Proceeding as in the earlier section, let $\alpha' = \alpha + \delta\alpha$, $\alpha'' = \alpha - \delta\alpha$, etc. and note that

$$S' = S(\alpha, \beta, \omega, z) + \frac{\partial S}{\partial \alpha}\delta\alpha + \frac{\partial S}{\partial \beta}\delta\beta + \frac{\partial S}{\partial \omega}\delta\omega + \ldots,$$

$$S'' = S(\alpha, \beta, \omega, z) - \frac{\partial S}{\partial \alpha}\delta\alpha - \frac{\partial S}{\partial \beta}\delta\beta - \frac{\partial S}{\partial \omega}\delta\omega + \ldots.$$

Then Eq. (64–2) becomes, on neglecting squares and higher powers of the increments,

$$\begin{aligned} P = 2 \sin (\alpha x + \beta y - \omega t + S) \cos \bigg[\left(x + \frac{\partial S}{\partial \alpha}\right) \delta\alpha \\ + \left(y + \frac{\partial S}{\partial \beta}\right) \delta\beta - \left(t - \frac{\partial S}{\partial \omega}\right) \delta\omega \bigg]. \end{aligned} \tag{3}$$

This may again be described as a modulated carrier wave, the carrier being determined by the three parameters α, β, ω, and the modulation by all six quantities α, β, ω, $\delta\alpha$, $\delta\beta$, $\delta\omega$. The argument continues as before: the modulation will have the constant magnitude

$$\cos (x_0\, \delta\alpha + y_0\, \delta\beta - t_0\, \delta\omega)$$

for all values of x, y, z, t that satisfy

$$\begin{aligned} x + \frac{\partial S}{\partial \alpha} &= x_0, = \text{const.}, \\ y + \frac{\partial S}{\partial \beta} &= y_0, = \text{const.}, \\ t - \frac{\partial S}{\partial \omega} &= t_0, = \text{const.} \end{aligned} \tag{4}$$

Equations (64–4), for constant α, β, ω, x_0, y_0, t_0, determine the motion of a point in space—this is the ray-point. Its path is now a curve, determined by the first two of Eqs. (64–4). Its position and shape depend on α, β, ω, x_0 and y_0. The third of Eqs. (64–4) determines the time at which the ray-point reaches a given altitude z.

The group velocity. The velocity with which the ray-point moves along the ray can be calculated as follows: its x-component is

$$c_x = (\mathrm{d}x/\mathrm{d}z)/(\mathrm{d}t/\mathrm{d}z)$$
$$= -\left[\frac{\partial}{\partial\alpha}\left(\frac{\partial S}{\partial z}\right)\right]\Big/\left[\frac{\partial}{\partial\omega}\left(\frac{\partial S}{\partial z}\right)\right]$$

by Eq. (64–4). Using Eq. (63–8), this is

$$c_x = -(\partial\gamma/\partial\alpha)/(\partial\gamma/\partial\omega)$$
$$= \frac{\alpha\omega c^2}{\omega^2 - \Omega^2} R, \tag{5}$$

where

$$R = \frac{\omega^2}{\omega^2 - N^2} - \varkappa^2 c^2 \frac{(N^2 - \Omega^2)\omega^2}{(\omega^2 - N^2)(\omega^2 - \Omega^2)^2}. \tag{6}$$

In the same way, the y- and z-components are

$$c_y = \frac{\beta\omega c^2}{\omega^2 - \Omega^2} R, \tag{7}$$

$$c_z = \frac{\gamma\omega c^2}{\omega^2 - N^2} R. \tag{8}$$

These equations are formally identical with Eqs. (51–5, 7); however, whereas c_x, c_y, c_z were independent of altitude in the case of the isothermal atmosphere, they now depend on altitude. The ray-point moves with the local value of the group velocity at the altitude to which it has attained.

Interpretation of the ray-point. There is another difference between these results and those of Section 51. The latter were derived rigorously for the isothermal atmosphere while the present results are approximate but applicable to any fluid. This corresponds to the fact that the W-K-B approximation becomes rigorous when the coefficients of the field equations are constants.

Within the limitations of the W-K-B approximation, however, the interpretation of the rays and the group velocity are the same as in Chapters VIII and IX. The rays are the stream-lines of the energy flow, and the group velocity is the velocity of the energy. Consequently, the ray-point can be considered to represent a small parcel of energy which moves along the ray with the group velocity.

Explicit equations for the rays. Using Eq. (63–9), Eqs. (64–4) become

$$x - x_0 = -\partial S/\partial\alpha = -\int (\partial\gamma/\partial\alpha)\,\mathrm{d}z$$

$$= -(\alpha/\varkappa)\int (\partial\gamma/\partial\varkappa)\,\mathrm{d}z$$

$$= \frac{\alpha}{\omega^2 - \Omega^2}\int [(\omega^2 - N^2)/\gamma]\,\mathrm{d}z, \tag{9}$$

$$y - y_0 = \frac{\beta}{\omega^2 - \Omega^2}\int [(\omega^2 - N^2)/\gamma]\,\mathrm{d}z, \tag{10}$$

$$t - t_0 = \omega\int \{[1/c^2 - \varkappa^2(N^2 - \omega^2)/(\omega^2 - \Omega^2)^2]/\gamma\}\,\mathrm{d}z. \tag{11}$$

General properties of the rays. Since N, N_1 and c are known functions of z, the determination of the rays and the travel time is therefore reduced to the evaluation of two integrals. Special cases of these integrals will be evaluated in the following chapters, and only some general conclusions will be noted here.

Equations (64–9, 10) show that

$$(x - x_0)/(y - y_0) = \alpha/\beta = \text{const.} \tag{12}$$

Hence the ray runs its entire course in a vertical plane. It will therefore be sufficient to treat the case $\alpha = \varkappa$, $\beta = 0$.

Since $x - x_0$, etc. must be real, it follows that the rays cannot penetrate those regions where

$$\gamma^2 < 0.$$

This is formally the same condition that led to the diagnostic diagram in Chapter VIII. Since γ now depends on z, one may consider that there is a diagnostic diagram for each altitude. However, it is more useful to consider α, β, ω as constants and z as the primary variable in γ.

Limiting forms of the rays. Equations (64–9, 10, 11) simplify in limiting cases. If N^2 and N_1^2 have upper bounds, the frequency ω can be chosen so large that the following are adequate approximations:

$$\begin{aligned} x - x_0 &= \varkappa \int \mathrm{d}z/[\omega^2/c^2 - \varkappa^2]^{\frac{1}{2}}, \\ t &= t_0 - \omega \int \mathrm{d}z/c[\omega^2/c^2 - \varkappa^2]^{\frac{1}{2}}. \end{aligned} \tag{13}$$

These are the familiar equations expressing Snell's Law for the refraction of sound rays. At those levels, z, for which

$$c(z) = \pm\omega/\varkappa,$$

the integrand becomes infinite, and the ray horizontal, $(\mathrm{d}z/\mathrm{d}x) = 0$. These are the vertices or turning points of the rays. The rays cannot pass into regions where $\omega^2 < \varkappa^2 c^2$, but are constrained to remain in the channels defined by $\omega^2 > \varkappa^2 c^2$. (See Fig. 52.)

If c also has a lower bound, $\varkappa^2$ can be chosen so much larger than $(\omega^2 - N_1^2)/c^2$ that the approximations

$$\begin{aligned} x - x_0 &= \int \left\{\frac{N^2 - \omega^2}{\omega^2 - \Omega^2}\right\}^{\frac{1}{2}} \mathrm{d}z, \\ t - t_0 &= -\frac{\omega\varkappa}{(\omega^2 - \Omega^2)^2} \int (N^2 - \Omega^2) \left\{\frac{\omega^2 - \Omega^2}{N^2 - \omega^2}\right\}^{\frac{1}{2}} \mathrm{d}z, \end{aligned} \tag{14}$$

are valid. These equations are both independent of c, and the first is independent of $\varkappa$. The shape of the rays is therefore independent of the wave-length and is determined solely by the frequency; the travel time is simply proportional to $\varkappa$. In order that the rays be real, $(N^2 - \omega^2)/(\omega^2 - \Omega^2)$ must be positive. If $N^2(z) > \Omega^2$ for all z, this is equivalent to $N^2 > \omega^2 > \Omega^2$; these are the rays of short gravity waves. At those levels determined by

$$N(z) = \pm\omega,$$

the integrand of the first equation becomes zero, and the ray is vertical, $(\mathrm{d}x/\mathrm{d}z) = 0$. The rays are reflected at these levels in a peculiar fashion, as shown by Fig. 40, and discussed more fully in Section 70.

65. Spherical level surfaces: no rotation

When the level surfaces are spheres, the considerations of the last section require some modification. This will first be discussed for the case $\Omega = 0$. The formulae of the Appendix then show that

$$\nabla\Phi = \frac{\cosh \mu}{r}\left(\boldsymbol{\epsilon}\frac{\partial\Phi}{\partial\lambda} + \boldsymbol{\nu}\frac{\partial\Phi}{\partial\mu}\right) + \boldsymbol{\zeta}\frac{\partial\Phi}{\partial r}.$$

Using $r = r_S + \chi$ as in Section 36, this is approximately

$$\nabla\Phi = \frac{\cosh \mu}{r_S}\left(\boldsymbol{\epsilon}\frac{\partial\Phi}{\partial\lambda} + \boldsymbol{\nu}\frac{\partial\Phi}{\partial\mu}\right) + \boldsymbol{\zeta}\frac{\partial\Phi}{\partial \chi}.$$

Using the definitions

$$\alpha = \frac{1}{r_S}\frac{\partial\Phi}{\partial\lambda}, \quad \beta = \frac{1}{r_S}\frac{\partial\Phi}{\partial\mu}, \quad \gamma = \frac{\partial\Phi}{\partial\chi}, \quad \omega = -\frac{\partial\Phi}{\partial t}, \tag{1}$$

(which are more appropriate to the present case than are Eqs. 62–3), this may be written

$$\mathbf{k} = \nabla\Phi = \cosh \mu(\alpha\boldsymbol{\epsilon} + \beta\boldsymbol{\nu}) + \gamma\boldsymbol{\zeta}. \tag{2}$$

The W-K-B approximation is still

$$\nabla P = i\mathbf{k}P, \quad \nabla Q = i\mathbf{k}Q, \nabla\cdot\mathbf{U} = i\mathbf{k}\cdot\mathbf{U}. \tag{3}$$

The approximate field equations are therefore

$$\begin{aligned} i\omega U_\epsilon &= i\alpha c \cosh \mu P, \\ i\omega U_\nu &= i\beta c \cosh \mu P, \\ i\omega U_\zeta &= (i\gamma + \Gamma)cP - N^2Q, \\ i\omega P &= ic \cosh \mu(\alpha U_\epsilon + \beta U_\nu) + (i\gamma - \Gamma)U_\zeta, \\ i\omega Q &= U_\zeta. \end{aligned} \tag{4}$$

After eliminating $\mathbf{U}$ from these, the residual equations are obtained in the approximate form

$$\begin{aligned} (i\gamma + \Gamma)cP - (N^2 - \omega^2)Q &= 0, \\ [1 - c^2(\alpha^2 + \beta^2)\cosh^2 \mu/\omega^2]P - (i\gamma - \Gamma)cQ &= 0. \end{aligned} \tag{5}$$

The determinant of these equations must vanish, yielding the propagation equation

$$(\gamma^2 + \Gamma^2)c^2 + (N^2 - \omega^2)[1 - c^2(\alpha^2 + \beta^2)\cosh^2\mu/\omega^2] = 0 \qquad (6)$$

Substitution from Eq. (65–1) then results in the Hamilton–Jacobi equation

$$c^2\left(\frac{\partial\Phi}{\partial t}\right)^2\left[\left(\frac{\partial\Phi}{\partial\chi}\right)^2 + \Gamma^2\right]$$

$$+\left[N^2 - \left(\frac{\partial\Phi}{\partial t}\right)^2\right]\left\{\left(\frac{\partial\Phi}{\partial t}\right)^2 - \frac{c^2\cosh^2\mu}{{r_S}^2}\left[\left(\frac{\partial\Phi}{\partial\lambda}\right)^2 + \left(\frac{\partial\Phi}{\partial\mu}\right)^2\right]\right\} = 0. \qquad (7)$$

The complete solution. The coefficients of this equation depend on χ and μ, but are independent of λ and t. Therefore it has solutions of the form

$$\Phi = V(\mu, \chi) + r_S\alpha\lambda - \omega t \qquad (8)$$

where α and ω are now constants. The function V must then satisfy the equation

$$\frac{\omega^2}{c^2(N^2-\omega^2)}\left[c^2\left(\frac{\partial V}{\partial\chi}\right)^2 + {N_1}^2 - \omega^2\right] = \frac{\cosh^2\mu}{{r_S}^2}\left[{r_S}^2\alpha^2 + \left(\frac{\partial V}{\partial\mu}\right)^2\right], \qquad (9)$$

obtained by substituting Eq. (25–8) into Eq. (25–7) and rearranging the terms. The variables in this equation are separated, in that the terms on the left are independent of μ, and those on the right are independent of χ. Consequently, it has solutions of the form

$$V(\mu, \chi) = S(\chi) + T(\mu), \qquad (10)$$

where

$$\left[\frac{1}{{r_S}^2}\left(\frac{dT}{d\mu}\right)^2 + \alpha^2\right]\cosh^2\mu = \varkappa^2, \qquad (11)$$

$$\frac{\omega^2}{c^2(N^2-\omega^2)}\left[c^2\left(\frac{dS}{d\chi}\right)^2 + {N_1}^2 - \omega^2\right] = \varkappa^2, \qquad (12)$$

and $\varkappa^2$ is the constant of separation. Solving these equations for the derivatives,

$$\frac{1}{r_S}\frac{dT}{d\mu} = \{\varkappa^2 \operatorname{sech}^2 \mu - \alpha^2\}^{\frac{1}{2}} = \beta(\alpha, \varkappa, \mu), \tag{13}$$

$$\frac{dS}{d\chi} = \{\varkappa^2(N^2 - \omega^2)/\omega^2 - (N_1^2 - \omega^2)/c^2\}^{\frac{1}{2}} = \gamma(\varkappa, \omega, \chi). \tag{14}$$

The last equation should be compared with Eq. (62–8). The explicit forms of T and S are then

$$T = r_S \int_{\mu_1}^{\mu} \beta(\varkappa, \alpha, \mu)\, d\mu, \tag{15}$$

$$S = \int_{\chi_1}^{\chi} \gamma(\varkappa, \omega, \mu)\, d\mu, \tag{16}$$

where μ_1 and χ_1 are again constants that may be chosen in any convenient way provided that T and S are real. Since this solution depends on the three non-additive constants $\varkappa$, α and ω, it is a complete one.

The rays. The equations for the rays can be derived by a repetition of the argument of Section 64. Without repeating this in detail, the result is that they are determined by the equations

$$\frac{\partial\Phi}{\partial\omega} = -t_0 = \text{const.},$$

$$\frac{\partial\Phi}{\partial\lambda} = r_S\lambda_0 = \text{const.},$$

$$\frac{\partial\Phi}{\partial\varkappa} = r_S\phi_0 = \text{const.},$$

which may be written more explicitly,

$$t - t_0 = \int \frac{\partial\gamma}{\partial\omega}\, d\chi, \tag{17}$$

$$\lambda - \lambda_0 = -\int \frac{\partial\beta}{\partial\alpha}\, d\mu, \tag{18}$$

$$r_S\phi_0 = r_S \int \frac{\partial\beta}{\partial\varkappa}\, d\mu + \int \frac{\partial\gamma}{\partial\varkappa}\, d\chi. \tag{19}$$

The rays are plane curves. Equation (65–18), when the explicit expression for β is substituted, reduces to the equation of a great circle (cf. Appendix, Eq. 16). In other words, the rays are plane curves, and their plane passes through the centre of the spherical level surfaces. This can be shown as follows: substitution from Eq. (65–13) converts Eq. (65–18) into

$$\lambda - \lambda_0 = \int_0^\mu \frac{\alpha \, d\mu}{\{\varkappa^2 \operatorname{sech}^2 \mu - \alpha^2\}^{\frac{1}{2}}} \tag{20}$$

provided the arbitrary lower limit of the integral is taken to be zero. Defining the quantity μ_0 by the equation

$$\cosh^2 \mu_0 = \varkappa^2/\alpha^2, \quad \mu_0 \geqslant 0, \tag{21}$$

Eq. (65–20) may be written

$$\begin{aligned} \lambda - \lambda_0 &= \int_0^\mu \frac{\cosh \mu \, d\mu}{\{\cosh^2 \mu_0 - \cosh^2 \mu\}^{\frac{1}{2}}} \\ &= \int_0^\mu \frac{d(\sinh \mu)}{\{\sinh^2 \mu_0 - \sinh^2 \mu\}^{\frac{1}{2}}}, \end{aligned} \tag{22}$$

which is easily integrated to yield

$$\sinh \mu = \sinh \mu_0 \sin (\lambda - \lambda_0). \tag{23}$$

This is the equation of a great circle and hence proves the result stated above.

If Eq. (65–23) is plotted on a Mercator chart of the planet's surface, the resulting curve is the central projection of the ray on the planetary surface; it is called the trace or track of the ray. Figure 34 shows the tracks for various values of μ_0. It is seen that they remain in the equatorial zone defined by $-\mu_0 \leqslant \mu \leqslant \mu_0$ and that they close on themselves: μ is a periodic function of λ, with period 2π.

Altitude of the rays. Equation (65–19) contains the integral

$$I = \int_0^\mu \frac{\partial \beta}{\partial \varkappa} d\mu = \int_0^\mu \frac{\varkappa \operatorname{sech}^2 \mu \, d\mu}{\{\varkappa^2 \operatorname{sech}^2 \mu - \alpha^2\}^{\frac{1}{2}}}. \tag{24}$$

This integral can be evaluated without difficulty, but its interpretation is rendered simpler by noting that the whole problem has spherical symmetry. It is therefore sufficient to consider only those rays whose tracks coincide with a meridian; for these, $\alpha = 0$, and the integral of Eq. (65–24) reduces to

$$I = \int_0^\mu \operatorname{sech} \mu \, d\mu = \phi, \tag{25}$$

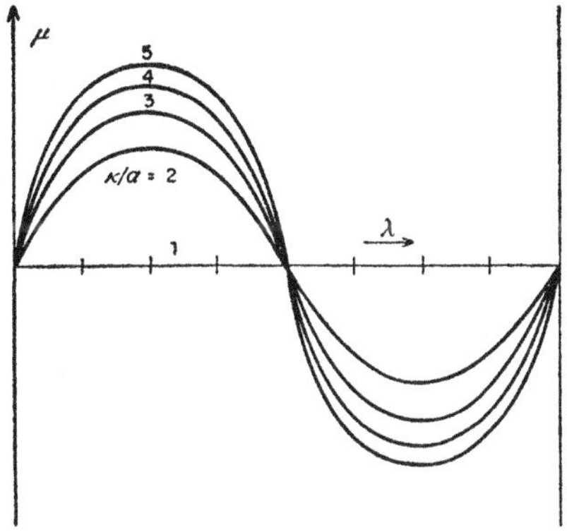

FIG. 34. Tracks of the rays, on a Mercator chart (no rotation).

where ϕ is defined by Eq. (2) of the Appendix and is therefore the latitude. Hence Eq. (65–19) becomes

$$r_S(\phi - \phi_0) = -\int_{\chi_1}^{\chi} \frac{\partial \gamma}{\partial \varkappa} \, d\chi, \tag{26}$$

which should be compared with Eq. (64–9). Even if the track of the ray is not a meridian, it follows from symmetry that the integral on the right of Eq. (65–26) gives the distance along its track as a function of the altitude of the ray.

66. Spherical level surfaces with rotation: traditional approximation

The influence of rotation on the rays will be investigated only with the traditional neglect of the horizontal component, Ω_H, of the

Coriolis vector. Using the notation of Eq. (65–1), the approximate field equations are

$$\begin{aligned}
i\omega U_\epsilon + \Omega \tanh \mu\, U_\nu &= i\alpha c \cosh \mu P,\\
i\omega U_\nu - \Omega \tanh \mu\, U_\epsilon &= i\beta c \cosh \mu P,\\
i\omega U_\zeta &= (i\gamma + \Gamma)cP - N^2 Q, \qquad (1)\\
i\omega P &= i\alpha \cosh \mu(\alpha U_\epsilon + \beta U_\nu) + (i\gamma - \Gamma)cU_\zeta,\\
i\omega Q &= U_\zeta.
\end{aligned}$$

Eliminating the components of the velocity field, the approximate residual equations are obtained:

$$(i\gamma + \Gamma)cP - (N^2 - \omega^2)Q = 0,$$

$$[1 - c^2(\alpha^2 + \beta^2)\cosh^2 \mu/(\omega^2 - \Omega^2 \tanh^2 \mu)]P \qquad (2)$$

$$-(i\gamma - \Gamma)cQ = 0.$$

The propagation equation is then

$$\gamma^2 c^2 = (\omega^2 - N_1{}^2)$$

$$+ (\alpha^2 + \beta^2)c^2 \cosh^2 \mu(N^2 - \omega^2)/(\omega^2 - \Omega^2 \tanh^2 \mu), \qquad (3)$$

which should be compared with Eq. (51–2).

Following an argument that is in all respects identical with that given in the preceding section, it is found that the phase function has the form

$$\Phi = S(\chi) + T(\mu) + r_S \alpha \lambda - \omega t, \qquad (4)$$

where

$$\left[\frac{1}{r_S{}^2}\left(\frac{dT}{d\mu}\right)^2 + \alpha^2\right]\frac{\omega^2 \cosh^2 \mu}{\omega^2 - \Omega^2 \tanh^2 \mu} = \varkappa^2, \qquad (5)$$

$$\frac{\omega^2}{c^2(N^2 - \omega^2)}\left[c^2\left(\frac{dS}{d\chi}\right)^2 + N_1{}^2 - \omega^2\right] = \varkappa^2, \qquad (6)$$

the constant of separation being $\varkappa^2$. Equation (66–5) reduces to Eq.

(65–11) when $\Omega = 0$, and Eq. (66–6) is the same as Eq. (65–12). Hence

$$\frac{1}{r_S}\frac{dT}{d\mu} = \{\varkappa^2 \operatorname{sech}^2 \mu[1 - (\Omega/\omega)^2 \tanh^2 \mu] - \alpha^2\}^{\frac{1}{2}}$$

$$= \beta(\alpha, \varkappa, \omega, \mu), \tag{7}$$

$$\frac{dS}{d\chi} = \gamma(\varkappa, \omega, \chi), \tag{8}$$

where γ is again given by Eq. (65–14) and does not depend on Ω. The integral forms for T and S are therefore

$$T = r_S \int_{\mu 1}^{\mu} \beta(\alpha, \varkappa, \omega, \mu)\, d\mu, \tag{9}$$

$$S = \int_{\chi 1}^{\chi} \gamma(\varkappa, \omega, \chi)\, d\chi, \tag{10}$$

and the equations of the rays are

$$t - t_0 = \int \frac{\partial\gamma}{\partial\omega}\, d\chi + r_s \int \frac{\partial\beta}{\partial\omega}\, d\mu, \tag{11}$$

$$\lambda - \lambda_0 = -\int \frac{\partial\beta}{\partial\alpha}\, d\mu, \tag{12}$$

$$s_0 = \int \frac{\partial\gamma}{\partial\varkappa}\, d\chi + r_S \int \frac{\partial\beta}{\partial\varkappa}\, d\mu, \tag{13}$$

t_0, λ_0, s_0 again being constants.

The tracks of the rays. Equation (66–12) again furnishes a relation between λ and μ, and yields the equation of the ray tracks on the planet's surface. These are no longer great circles, however. Explicitly, Eq. (66–12) is

$$\lambda - \lambda_0 = \int_0^{\mu} \frac{\alpha\, d\mu}{\{\varkappa^2 \operatorname{sech}^2 \mu[1 - (\Omega(\omega)^2 \tanh^2 \mu] - \alpha^2\}^{\frac{1}{2}}}\,. \tag{14}$$

The change of variable

$$\tau = \tanh \mu, \quad d\tau/d\mu = \operatorname{sech}^2 \mu = 1 - \tau^2, \tag{15}$$

transforms it into

$$\lambda - \lambda_0 = \int_0^\tau \frac{d\tau}{(1 - \tau^2)F} \tag{16}$$

where

$$F = \{(\varkappa/\alpha)^2(1 - \tau^2)[1 - (\Omega/\omega)^2\tau^2] - 1\}^{\frac{1}{2}}. \tag{17}$$

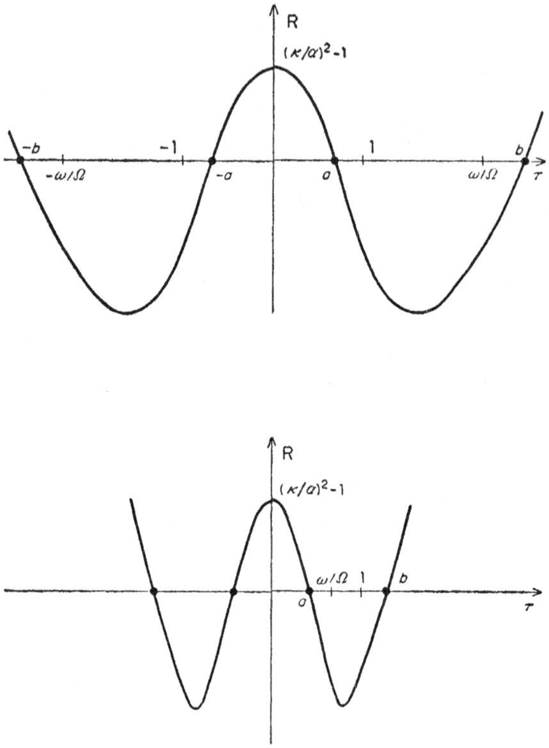

FIG. 35. Illustrating the roots of the function F.

This is an elliptic integral of the third kind, and while it can be evaluated with the aid of tables,[9] it will here be sufficient to derive certain qualitative facts concerning it.

The radicand of F is a polynomial of second degree in τ^2; its graph is shown on Fig. 35 for the two cases $\omega > \Omega$ and $\omega < \Omega$. It has four roots, $\tau = \pm a, \pm b$; it is seen that a is always less than unity and that F^2 is positive when $-a \leqslant \tau \leqslant a$. Hence the ray is confined to this range of τ. The polynomial is also positive for $\tau > b > 1$; but since $\tau = \tanh \mu$, this range does not lead to physically possible rays.

Concerning the approximation. It is also seen that even when $\omega < \Omega$, $a < \omega/\Omega$. If $\varkappa^2/\alpha^2$ is not too large, a will even be considerably less than ω/Ω. Under these conditions, the whole course of the ray will be confined to a region where $\omega \ll \Omega\tau = \Omega \tanh \mu = \Omega_V$. Now, in Chapter IX, it was seen that the errors caused by the traditional approximation increase as ω approaches Ω_V. It is therefore interesting to note that there is some tendency for the rays to avoid the region where the errors become very large. This does not, of course, exclude the possibility that the errors are appreciable.

The period of the ray tracks. Supposing the integral in Eq. (66–16) to have been evaluated, it would express $\lambda - \lambda_0$ as a function of $\tau = \tanh \mu$. This could be inverted into the form

$$\sinh \mu = f(\lambda - \lambda_0). \tag{18}$$

In the case $\Omega = 0$, this equation must reduce to Eq. (65–23):

$$\sinh \mu = \sinh \mu_0 \sin (\lambda - \lambda_0).$$

In that case f is a periodic function of its argument, and its period is 2π. In the general case, $\Omega \neq 0$, f remains a periodic function, but its period is not 2π. Instead, it can be shown to have the period

$$\Lambda = 4 \int_0^a \frac{d\tau}{(1 - \tau^2)F}. \tag{19}$$

For large values of ω/Ω, Λ approaches 2π, the difference $\Lambda - 2\pi$ being proportional to $(\Omega/\omega)^2$. For small values of ω/Ω, on the other hand, Eq. (66–14) may be approximated as

$$\lambda - \lambda_0 = \int_0^\mu \frac{\alpha\, d\mu}{\{\varkappa^2(1 - \Omega^2\mu^2/\omega^2) - 1\}^{\frac{1}{2}}}, \tag{20}$$

which integrates to

$$\mu = \frac{\omega}{\Omega}(1 - \alpha^2/\varkappa^2)^{\frac{1}{2}} \sin \left[\frac{\varkappa\Omega}{\alpha\omega}(\lambda - \lambda_0)\right]. \tag{21}$$

In this limit, the period is thus

$$\Lambda = 2\pi(\alpha\omega/\varkappa\Omega), \tag{22}$$

and is much less than 2π. The track of the ray thus crosses the equator at the longitudes

$$\lambda_n = \lambda_0 + n\pi(\alpha\omega/\varkappa\Omega). \tag{23}$$

Altitude of the rays. Equation (66–13) furnishes a relation between the altitude of the rays and the Mercator co-ordinate μ. In discussing the corresponding relation in Section 65, considerable simplication was possible because of the spherical symmetry of the problem. The rotation destroys this symmetry, and Eq. (66–13) is therefore considerably more complicated than the previous Eq. (65–19).

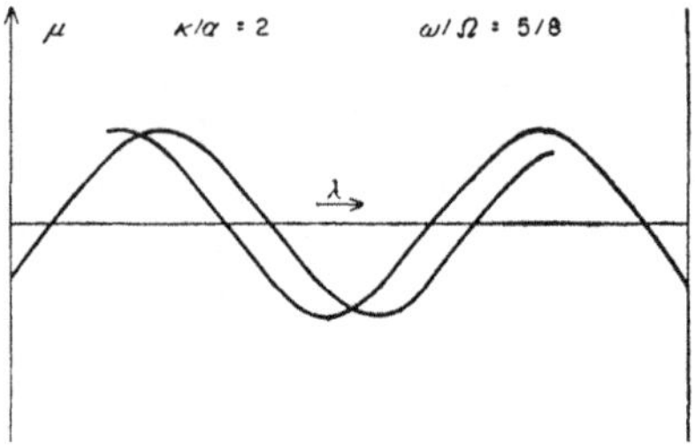

FIG. 36. Tracks of the rays, on a Mercator chart (with rotation).

The co-ordinate μ enters Eq. (66–13) through the integral

$$I = r_S \int \frac{\partial\beta}{\partial\varkappa}\, \mathrm{d}\mu$$

$$= r_S \int \frac{\varkappa \operatorname{sech}^2 \mu[1 - (\Omega/\omega)^2 \tanh^2 \mu]}{\beta(\alpha, \varkappa, \omega, \mu)}\, \mathrm{d}\mu. \tag{24}$$

This integral will now be compared with the integral that determines the distance along the track of the ray; in Section 65, where Ω was zero, the two corresponding integrals were found to be equal.

Using Eq. 4 of the Appendix, the element of length, ds, is given by

$$(\mathrm{d}s)^2 = r_S{}^2 \operatorname{sech}^2 \mu(\mathrm{d}\lambda^2 + \mathrm{d}\mu^2).$$

From Eq. (66–12)

$$\mathrm{d}\lambda = -\frac{\partial\beta}{\partial\alpha}\, \mathrm{d}\mu,$$

so that

$$(ds)^2 = r_S^2 \operatorname{sech}^2 \mu \left[\left(\frac{\partial \beta}{\partial \alpha} \right)^2 + 1 \right] d\mu^2$$

$$= r_S^2 \operatorname{sech}^2 \mu \left[\frac{\alpha^2}{\beta^2} + 1 \right] d\mu^2. \qquad (25)$$

Using the Eq. (66–7) for β, it is readily found that

$$ds = r_S \frac{\varkappa \operatorname{sech}^2 \mu [1 - (\Omega/\omega^2) \tanh^2 \mu]^{\frac{1}{2}}}{\beta(\alpha, \varkappa, \omega, \mu)} d\mu. \qquad (26)$$

It follows that

$$I = \int [1 - (\Omega/\omega)^2 \tanh^2 \mu]^{\frac{1}{2}} ds. \qquad (27)$$

If $\Omega = 0$, this reduces to $I = s$, and the result of the previous section is recovered. In the general case, it is seen that $I < s$. Hence changes in the altitude of the ray occur more slowly along the track than they would if there were no rotation, and this is especially pronounced for small values of ω/Ω.

References for Chapter XI

1. H. Bateman, Partial differential equations of mathematical physics, 1st American ed., Dover, New York, 522 pp., 1944.
2. R. Bellman and R. Kalaba, Invariant imbedding, wave propagation, and the WKB Approximation, *Proc. Natl. Acad. Sci., U.S.* **44,** p. 317 (1958).
3. L. Bieberbach, *Theorie der Differentialgleichungen,* 3rd ed., Dover, New York, 1944.
4. H. Bremmer, The WKB Approximation as the first term of a geometric-optical series, in *Theory of Electromagnetic Waves—A Symposium,* Interscience, New York, 1951.
5. L. Brillouin, *J. de Physique* **7,** p. 353 (1926).
6. C. Eckart, Note on the correspondence principle in the new quantum theory, *Proc. Natl. Acad. Sciences, U.S.* **12,** 685 (1926).
7. C. Eckart, Die Korrespondenzmassige Beziehung zwischen den Matrizen und den Fourierkoeffizienten des Wasserstoffs, *Z. für Physik* **48,** 295 (1928).
8. H. A. Kramers, *Z. für Physik* **39,** 829 (1926).
9. L. M. Milne-Thompson, *Jacobian Elliptic Functions,* Dover, New York, 1950.
10. Frank and von Mises (ed.), *Riemann-Waber's Partiellen Differentialgleichungen,* Vieweg, Braunschweig, 1925.
11. A. G. Webster, *Partial Differential Equations of Mathematical Physics,* Tenbner, Leipzig, 440 pp., 1927.
12. G. Wentzel, *Z. für Physik* **38,** 518 (1926).

CHAPTER XII

THE THERMOCLINE

67. Formulation of the problem

THE most striking characteristic of the Earth's oceans is the permanent thermocline. This region of rapidly changing temperature is also a region of great stability and results in the pronounced maximum of the parameter N, shown on Fig. 10 of Section 27. This chapter will be devoted to a study of the implications of this type of stratification.

For this purpose, it will be supposed that N is given by the simple analytic formula of Eq. (67–1) below. There is little doubt that this formula deviates considerably from the empirical curve of N as a function of depth (compare Figs. 10 and 37), but the two curves both have the characteristic maximum of the thermocline. Equation (67–1) has the advantage that many calculations can be made analytically, whereas the use of the empirical relation would require a resort to numerical computations. Moreover, many of the qualitative results will be independent of the particular analytic form adopted and depend only on the existence of the single maximum of N at the thermocline. To simplify the notation, the origin of z will be taken above the free surface, whose vertical co-ordinate will then be $z = -S$. That of the bottom will be $z = -S - D$, and the formula for N will be supposed to be

$$N^2 = -N_M{}^2[(2Z/z) + (Z^2/z^2)], \qquad (1)$$

whose graph is shown by Fig. 37. Here N_M and Z are constants; the former is the maximum value of the stability parameter, and Z is the thermocline depth. Obviously, only the range $-S - D \leqslant z \leqslant -S$ will be of physical significance. Since boundary conditions are ignored in the theory of simple waves, the parameters S and D will be of minor importance in the early part of this chapter, and the whole range $-\infty \leqslant z \leqslant 0$ will be drawn into the ray calculations. Clearly the "depth" $z = 0$ will be singular, and this singularity will be without physical significance. It will also be noted that, at both ends of this range, the stability will be subcritical (see Section 52) and even negative near $z = 0$; this is also without physical significance. It may be assumed that the subcritical ranges do not extend into the range of z-values that is actually occupied by the water.

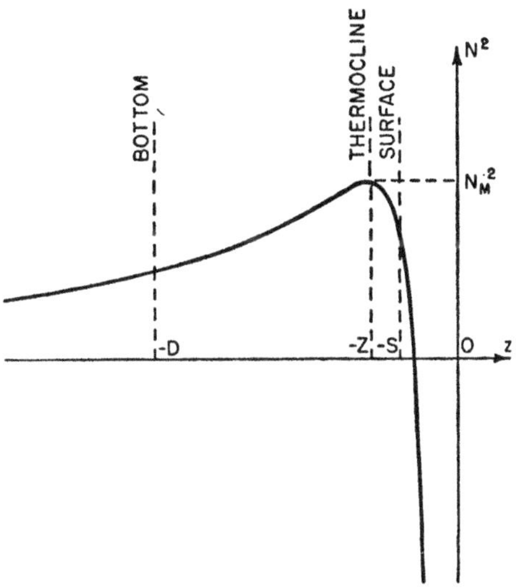

FIG. 37. Variation of N^2 with depth, as assumed for the illustrative purposes of this chapter.

Further simplification can be achieved by setting $\Gamma = 0$ and supposing that the velocity of sound, c, is constant. The former will have only negligible quantitative consequences; the latter implies that the rays of high-frequency sound will be straight lines. In the actual ocean the refraction of high-frequency sound is often of importance but would be an unnecessary complication here. At present, the primary objective is a study of the internal gravity waves associated with the thermocline, and these are scarcely affected by the velocity of sound, as is made evident by the results of Chapter X.

68. Preliminary discussion of the rays

At first, the analysis will be based on ray theory even though this can only yield approximate results. These will later be compared with results derived from the residual equations. For those rays that lie in the (x,z) plane, $\alpha = \varkappa$ and $\beta = 0$, so that Eq. (64–9) is

$$x - x_0 = -\int_{z_1}^{z} \frac{\partial \gamma}{\partial \varkappa}\, \mathrm{d}z. \tag{1}$$

Since $\Gamma = 0$, Eq. (63–8) reduces to

$$\gamma = \frac{1}{c}\{(\omega^2 - N^2)[1 - \varkappa^2 c^2/(\omega^2 - \Omega^2)]\}^{\frac{1}{2}}, \tag{2}$$

so that Eq. (68–1) is, in explicit form

$$x - x_0 = \varkappa c \int_{z_1}^{z} \left\{\frac{\omega^2 - N^2}{(\omega^2 - \Omega^2)(\omega^2 - \Omega^2 - \varkappa^2 c^2)}\right\}^{\frac{1}{2}} \mathrm{d}z. \tag{3}$$

As always, the radicand must be positive over the whole range of integration from z_1 to z, since the co-ordinates x and x_0 would otherwise be complex numbers.

Since c is a constant, the denominator of the radicand is also constant, and only the numerator depends on z. However, the sign will be affected by both numerator and denominator. The latter changes sign on the hyperbola

$$\omega^2 = \Omega^2 + \varkappa^2 c^2$$

and on the line

$$\omega = \Omega$$

both of which are plotted on Fig. 38. These, together with the line

$$\omega = N_M$$

divide the $(\varkappa,\omega)$ diagram into five areas, as shown on Fig. 38. It is convenient to discuss each of these five areas separately. The general correspondence between Fig. 38 and Fig. 21 of Section 50 should be noted.

Area I

Here $\omega^2 > N^2$ for all negative values of z, so that $\omega^2 - N^2 \geqslant 0$. The factor $\omega^2 - \Omega^2$ is also positive, but $\omega^2 - \Omega^2 - \varkappa^2 c^2$ is negative; the radical in Eq. (68–2) is therefore imaginary for all values of z, and there are no rays associated to points in this area of the diagnostic.

AREA II

Here all three factors are positive for all negative z. The rays are real and may extend from $-\infty$ to 0. These rays correspond to acoustic waves, as will become evident below, if it is not already clear.

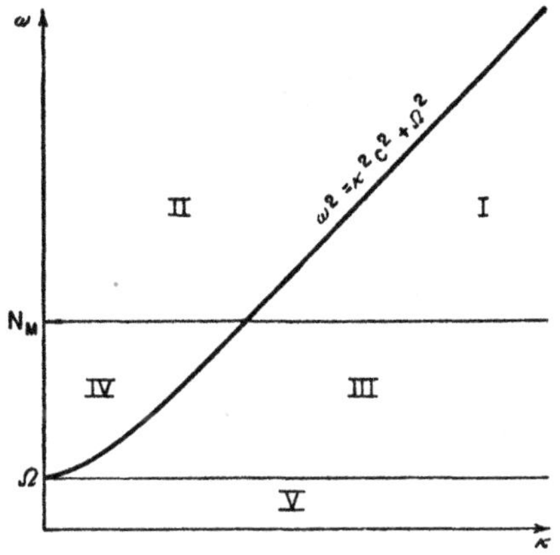

FIG. 38. The five special cases to be considered, showing the location of the $\varkappa,\omega$ points in the diagnostic diagram.

AREA III

Here the denominator of the radicand is negative; the rays are therefore confined to the region where $N^2 > \omega^2$. Let $z_1$and z_2 be the roots of

$$N^2(z) = \omega^2, \qquad z_1 \leqslant z_2.$$

Then z_1 is conveniently taken as the lower limit of the integral in Eq. (68–3), and the rays are confined to the range $+z_1 \leqslant z \leqslant z_2$. At the two ends of this range,

$$\frac{dx}{dz} = \gamma = 0,$$

and the rays are vertical. These rays correspond to the internal gravity waves. Since these rays are confined to the region of the thermocline, one may speak of the thermocline-channel for internal waves.

AREAS IV AND V

The denominator of the radicand is positive in both these areas, and the radical will be real outside the segment defined by the roots of

$N^2 = \omega^2$. There are two channels. In the lower one, the rays will be found to come from $z = -\infty$ and to turn, with vertical tangent, at the lower root before reaching the thermocline; they then recede to $-\infty$ again. The other channel is above the upper root and extends to $z = 0$. Of course, the rays will reach the bottom or surface and be reflected there according to the simple law of equal angles. Consequently the two channels may be called the bottom- and the surface-channel, respectively. The exact nature of these waves cannot be foreseen at this stage of the discussion.

This preliminary analysis indicates that three cases will have to be distinguished in evaluating the integral of Eq. (68–1). These correspond respectively to area II, area III and areas IV and V.

69. The sound waves of area II

When $\varkappa,\omega$ is a point of area II, Eq. (68–3) may be written

$$\begin{aligned} x - x_0 &= AZ\int \frac{(a^2 - 2u + u^2)^{\frac{1}{2}}}{u^2}\,du \\ &= AZ\int \frac{[b^2 + (u-1)^2]^{\frac{1}{2}}}{u^2}\,du \end{aligned} \tag{1}$$

where

$$\begin{aligned} u &= -Z/z, \\ a = \omega/N_M > 1, &\qquad b = (a^2 - 1)^{\frac{1}{2}}, \\ A &= \varkappa c N_M[(\omega^2 - \Omega^2)(\omega^2 - \Omega^2 - \varkappa^2 c^2)]^{-\frac{1}{2}}. \end{aligned} \tag{2}$$

The integrand is real for all real u. The case of very large a is of importance since it corresponds to sound waves of audio frequency. The shape of these rays near the thermocline will first be investigated. Then $(u - 1)/b$ will be a small quantity, and

$$(a^2 - 2u + u^2)^{\frac{1}{2}} = b\{1 + (u - 1)^2/2b + \ldots\}.$$

It follows that

$$x - x_0 = Ab\{z + (Z/2b^2)[(z/Z) + 2\log(-z/Z)\ldots\}, \tag{3}$$

so that the rays depart only little from straight lines of slope Ab. To a certain approximation, this slope is $\omega/(\omega^2 - \varkappa^2c^2)^{\frac{1}{2}}$, which is the classic formula for sound rays. However, the departures from a straight line are of theoretical interest since they indicate a small refraction of sound rays, despite the fact that the velocity of sound c is constant.

At points far beneath the thermocline, u becomes very small, and

$$(a^2 - 2u + u^2)^{\frac{1}{2}} = a\left\{1 - \frac{u}{a^2} + \frac{1}{2}\frac{u^2}{a^2}\left(1 - \frac{1}{2a^2}\right)\ldots\right\},$$

so that

$$x - x_0 = Aa\left\{z + \frac{Z}{a^2}\log(-z/Z) + \ldots\right\}. \tag{4}$$

Consequently, there are departures from a straight line even at these depths. Moreover, the slopes at the thermocline and at great depths

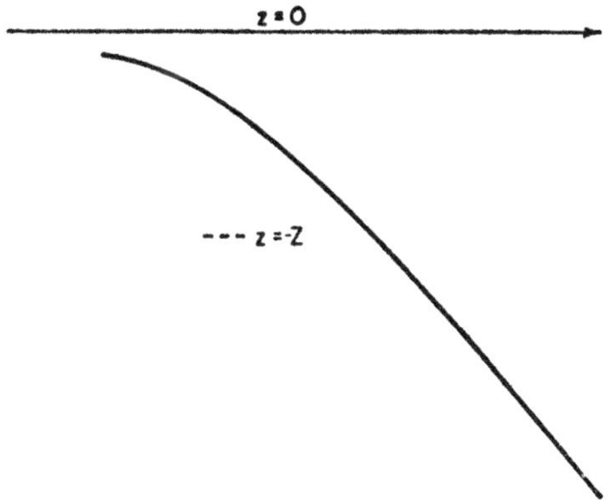

FIG. 39. A sound ray of AREA II.

are not equal; their ratio differs from unity by amounts of the order $(N_M/\omega)^4$. This will normally be much below the limit of experimental verification.

Large values of u (small z) are without physical significance, but formally Eq. (69–1) leads to

$$x - x_0 = AZ\log(-Z/z)\ldots \tag{5}$$

in this case. The large negative values of N^2 would cause a very appreciable refraction of the sound rays, but are not physically realizable.

The integral of Eq. (69–1) can also be evaluated without approximation: in terms of the parameters ϕ and ψ defined by

$$u = 1 + b\sinh\phi = a^2/[1 + b\sinh\psi] = -Z/z, \tag{6}$$

the integral of Eq. (69–1) may be evaluated and is

$$x - x_0 = AZ[\phi - (\psi + b \cosh \psi)/a]. \tag{7}$$

Figure 39 shows the graph of this equation; the numerical values used in constructing this figure have been chosen for clarity and not for realism.

70. The gravity waves of area III

When $\varkappa,\omega$ is a point of area III, Eq. (68–3) may be written

$$(x - x_0) = AZ\int \frac{(-a^2 + 2u - u^2)^{\frac{1}{2}}}{u^2}\, du \tag{1}$$

where now

$$u = -Z/z,$$
$$a = \omega/N_M < 1, \qquad b = (1 - a^2)^{\frac{1}{2}} \tag{2}$$
$$A = \varkappa c N_M[(\omega^2 - \Omega^2)(\varkappa^2 c^2 + \Omega^2 - \omega^2)]^{\frac{1}{2}}.$$

The integrand has roots at $u = 1 \pm b$ and is real only when u is between these roots. Its behaviour near the root $u = 1 + b$ may be investigated by introducing the variable

$$v = 1 + b - u, \tag{3}$$

so that

$$x - x_0 = \pm AZ\int \frac{[v(2b - v)]^{\frac{1}{2}}}{(1 + b - v)^2}\, dv \tag{4}$$

The arbitrary sign of the radical has hitherto been ignored but must now be drawn into the calculation. Treating v as a small quantity, the integral of Eq. (70–4) becomes

$$x - x_0 = \pm AZb^{\frac{1}{2}}[(2v)^{\frac{3}{2}}/3(1 + b)^2] + \dots. \tag{5}$$

The ray is therefore reflected at the level $z/Z = -1/(1 + b)$ and has a vertical tangent at that level. The behaviour of the ray at the other root is similar.

This peculiar behaviour of the ray near these roots is closely related to the propagation surfaces discussed in Section 43. Since N is a function of z, there will be such a surface—the local propagation surface—associated with each point of the ray. The parameters $\varkappa$ and ω are the same for all the local surfaces, but γ and N vary with z.

Throughout the course of these rays, $\omega^2 > N^2$, so that the propagation surface is always an hyperboloid, as shown on the lower part of Fig. 17. When z approaches one of the roots of $N^2 = \omega^2$, the asymptotic cone of the hyperboloid flattens, and the inclination of the ray becomes more and more nearly vertical as will be clear from a study of Fig. 17.

The integral of Eq. (70–1) is again rigorously expressible in terms of ϕ and ψ, which are now defined by

$$u = 1 + b \sin \phi = a^2/(1 + b \sin \psi) = -Z/z. \tag{6}$$

It is

$$x - x_0 = -AZ[\phi + (\psi + b \cos \psi)/a]. \tag{7}$$

Figure 40 shows its graph.

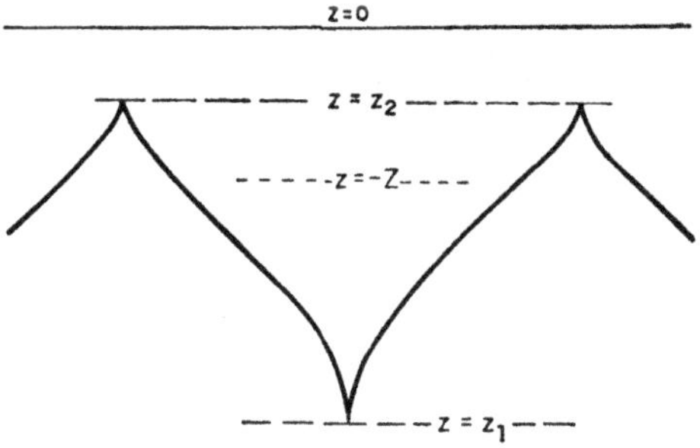

FIG. 40. The ray of a gravity wave of AREA III.

It may be, of course, that the surface or bottom intersects the rays. This will occur for sufficiently small values of ω provided that the stability is not sub-critical at any depth. In this case, the ray will be reflected at the interface according to the law of equal angles. If the rays are not reflected at either interface, they run their entire course in the thermocline-channel.

71. The waves of areas IV and V

The analytic theories of areas IV and V are identical; however there are physical differences that must be considered afterwards. Equation (68–3) may be written

$$x - x_0 = AZ \int \frac{(a^2 - 2u + u^2)^{\frac{1}{2}}}{u^2} \, du \tag{1}$$

if the definitions

$$u = -Z/z,$$

$$a = \omega/N_M < 1, \qquad b = (1 - a^2)^{\frac{1}{2}},$$

$$A = \varkappa c N_M[(\omega^2 - \Omega^2)(\omega^2 - \Omega^2 - \varkappa^2 c^2)]^{-\frac{1}{2}} \tag{2}$$

are used. The integrand of Eq. (71–1) is now real outside of the segment between its two roots, and two cases must be distinguished

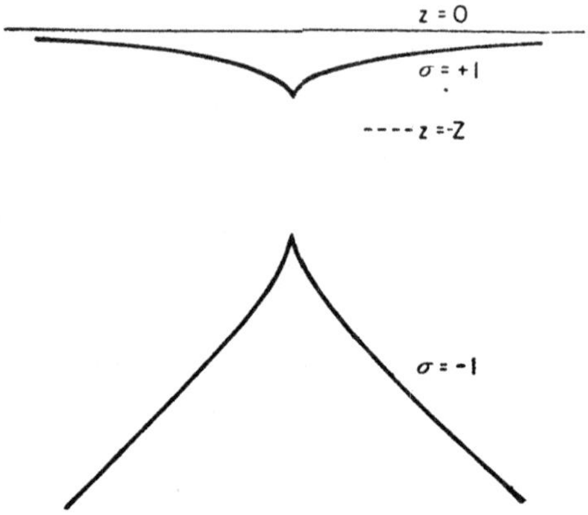

FIG. 41. Rays of AREAS IV and V.

by a parameter $\sigma = \pm 1$. If u is greater than either root, then $\sigma = +1$; if u is less than either root, then $\sigma = -1$. The definitions of ϕ and ψ are now

$$u = 1 + \sigma b \cosh \phi = a^2/[1 - \sigma b \cosh \psi] = -Z/z, \tag{3}$$

and

$$x - x_0 = AZ\{\sigma\phi - [\sigma\psi + b \sinh \psi]/a\}. \tag{4}$$

The graphs of these rays are shown by Fig. 41.

In area V, $\omega < \Omega$, and hence the rays are confined to the sub-critical range where $N < \Omega$. If these do not extend into the actual water, these rays are entirely fictitious, and there are no solutions for $\omega < \Omega$. This remark emphasizes the importance of determining whether the stability of the oceans becomes sub-critical at great depths.

In area IV, $\omega > \Omega$, and hence either branch of the ray can extend into the actual water even though the stability is not sub-critical in the range of physically significant values of z. However, if they do, then the surface or bottom will intersect the rays, and they will be reflected there. The rays are thus confined to the surface or bottom channel as the case may be.

72. The residual equations

Having surveyed the problem of the thermocline, using the theory of rays, it will now be examined in terms of the residual equations. With $\Gamma = 0$, these are

$$c\frac{\mathrm{d}P}{\mathrm{d}z} = (N^2 - \omega^2)Q, \qquad c\frac{\mathrm{d}Q}{\mathrm{d}z} = [1 - \varkappa^2c^2/(\omega^2 - \Omega^2)]P. \tag{1}$$

The boundary conditions are

$$Q = 0 \quad \text{at} \quad z = -S - D \text{ (bottom)}, \tag{2}$$

$$cP + gQ = 0 \quad \text{at} \quad z = -S \text{ (surface)}. \tag{3}$$

If the field P is eliminated from these equations, Eqs. (72–1) yield, because c is constant

$$\frac{\mathrm{d}^2Q}{\mathrm{d}z^2} + \gamma^2Q = 0, \tag{4}$$

where γ is again given by Eq. (68–2):

$$\gamma^2c^2 = \left(1 - \frac{\varkappa^2c^2}{\omega^2 - \Omega^2}\right)(\omega^2 - N^2). \tag{5}$$

Equation (72–2) remains unaltered while Eq. (72–3) becomes

$$\frac{\mathrm{d}Q}{\mathrm{d}z} = -\frac{g}{c^2}\left[1 - \frac{\varkappa^2c^2}{\omega^2 - \Omega^2}\right]Q \quad \text{at} \quad z = -S. \tag{6}$$

It will be recalled that Q is essentially the amplitude of the vertical displacement of the water, measured in the appropriate units.

It will next be shown that the solutions of Eq. (72–4) have different characteristics in each of the five areas of Fig. 38 and that the roots of $N^2 = \omega^2$ again are important parameters. In other words, the theory of the residual equations parallels the ray theory in these important respects.

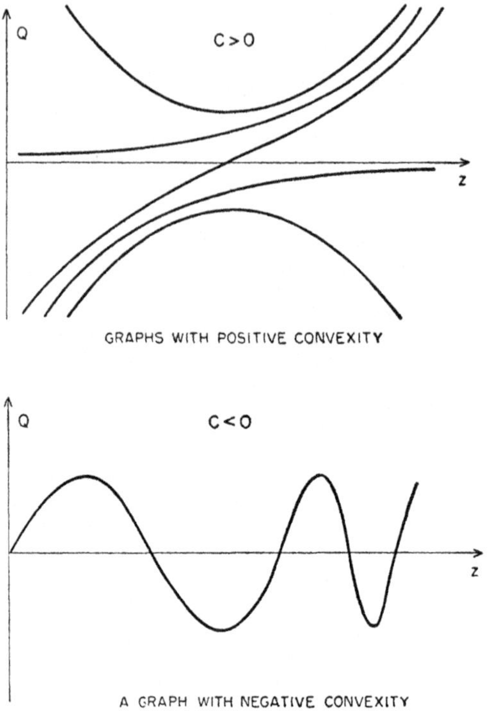

FIG. 42. Typical graphs with positive or negative convexity.

The solution of Eq. (72–4) can be represented graphically by plotting the values of Q along the z-axis. An important geometric property of this graph is its convexity, C, which is defined by

$$C = \frac{1}{Q}\frac{\mathrm{d}^2Q}{\mathrm{d}z^2}\,. \tag{7}$$

When the convexity is positive, the graph is convex toward the z-axis; when $C < 0$, it is convex away from the z-axis. These two cases are illustrated on Fig. 42.

Consider an interval of the z-axis over which C does not change sign. If the convexity is positive throughout this interval then Q can have at most one root or one extremum (but not both), as is shown by the upper part of Fig. 42. If, on the other hand, the convexity is negative over the interval, then Q may have many roots and extrema, as shown by the lower part of Fig. 42.

If $C > 0$ and Q has a minimum, that minimum value will be positive; if $C < 0$, the minima are negative. Correspondingly, the maxima of Q have a sign that is opposite to that of C. It follows that, when $C < 0$, the extrema of Q are separated by roots; consequently, is said to oscillate in those intervals where $C < 0$. If $C \neq 0$, the roots of Q are also inflexion points of its graph; the roots of C are inflexion points, but not roots, of Q.

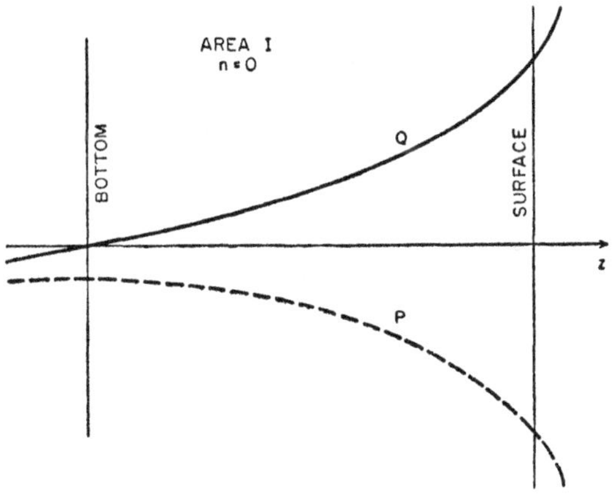

FIG. 43. Sketch of the eigensolution for $n = 0$ and $\varkappa, \omega$ in AREA I.

These simple remarks make it possible to construct fairly detailed though qualitative graphs of the solutions of Eq. (72–4) without extensive numerical calculations

AREA I

This qualitative construction will now be worked out for the various areas of the diagnostic diagram, Fig. 38. In area I,

$$\Omega^2 < N_M{}^2 < \omega^2 < \varkappa^2 c^2 + \Omega^2. \tag{8}$$

Hence $\gamma^2 < 0$, and the convexity of the Q graph is positive for all values of z. Equation (72–2) shows that it must have a root at $z = -S - D$; it can therefore have no other root. A sketch of the graph must therefore be somewhat as shown on Fig. 43, provided that $\mathrm{d}Q/\mathrm{d}z > 0$ at the bottom.

Equations (72–2) and (72–4) do not determine the value of this slope which is therefore an arbitrary parameter in the solution. However, the value of

$$\frac{1}{Q}\frac{dQ}{dz} = f \tag{9}$$

is uniquely determined by these equations and is a function of $\varkappa$ and ω at each point: $f = f_0(\varkappa, \omega, z)$. While the calculation of this function would require extensive work, it is determined uniquely; moreover, it is obvious that $f_0 > 0$ for $z > -S - D$.

Substitution into Eq. (72–6) then yields the equation

$$f_0(\varkappa, \omega, -S) = \frac{g}{c^2}\left[\frac{\varkappa^2 c^2}{\omega^2 - \Omega^2} - 1\right] \tag{10}$$

connecting $\varkappa$ and ω; this should be compared with the discussion in Section 54.

Finally, Eq. (72–1) shows that P is, except for a constant negative factor, equal to dQ/dz, so that the graph of P must be essentially as shown on Fig. 43. The fields P and Q are thus 180° out of phase, both reach their greatest amplitude at the surface, and P has no node. These are all characteristics of the surface wave mode, $n = 0$, as discussed in Chapter X.

Area II

It is to be presumed that $\varkappa,\omega$ points in area II will lead to the acoustic modes. Here

$$\Omega^2 < N_M{}^2 < \varkappa^2 c^2 + \Omega^2 < \omega^2, \tag{11}$$

so that $\gamma^2 > 0$ and the convexity is negative over the whole z-axis. The graph of Q will thus be oscillatory, but, as before, Eq. (72–2) requires that it have a node at $z = -S - D$, as shown on Fig. 44. There may also be other nodes in the range $-S - D < z < -S$; let n be the total number of these. The function f, of Eq. (72–9), is again uniquely determined but will be a different function of $\varkappa,\omega$ and $z : f = f_n(\varkappa, \omega, z)$. The surface boundary condition requires that

$$f_n(\varkappa, \omega, z) = \frac{g}{c^2}\left[\frac{\varkappa^2 c^2}{\omega^2 - \Omega^2} - 1\right]. \tag{12}$$

Since the second term of Eq. (72–12) is negative, Q and dQ/dz must have opposite signs at the surface. The pressure field is now proportional to dQ/dz, with a positive factor, and therefore also has n nodes in the physically significant range of z-values. These are all characteristics of the nth acoustic mode.

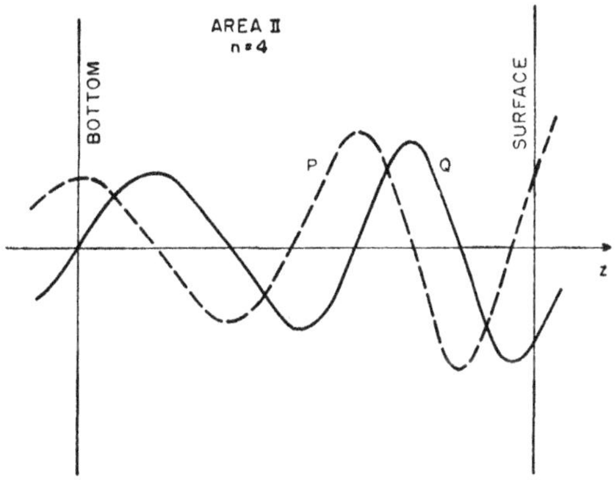

FIG. 44. Sketch of the eigensolution for $n = 4$ and $\varkappa,\omega$ in AREA II.

AREA III

This is perhaps the most interesting case since it should yield the internal gravity waves and also show how the thermocline modifies the low-frequency surface waves. In this area

$$\Omega^2 < \omega^2 < N_M{}^2 \text{ and } \omega^2 < \varkappa^2 c^2 + \Omega^2. \tag{13}$$

Hence the equation

$$N^2 = \omega^2 \tag{14}$$

has two roots, $z = -z_1$ and $z = -z_2$, and γ^2 changes sign. It is readily seen that the convexity of Q satisfies the inequalities

$$C > 0 \quad \text{when} \quad z < -z_1, \text{ or } z > -z_2,$$

$$C < 0 \quad \text{when} \quad -z_1 < z < -z_2.$$

It is also seen that, as ω decreases, the separation of the roots increases, as does the general magnitude of the convexity. As a consequence of these matters, the graphs of Q and P are less simple than in the cases considered above but can still be sketched by applying the same principles. Figures 45, 46 and 47 show three cases in which P has

0, 1, 2 nodes and which, according to the convention established in Chapter X, have the ordinal numbers $n = 0, -1, -2$.

$n = 0$.

In this case, Q has only the node at the bottom, $z = -S - D$. Its graph is convex away from the z-axis in the range between $-z_1$ and $-z_2$, convex toward the z-axis outside this range. However, its slope does not change in sign. The ratio of P to dQ/dz is negative

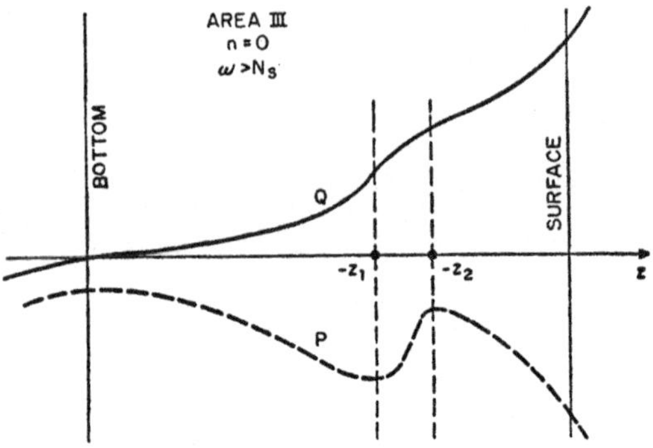

FIG. 45. Sketch of the eigensolution for $n = 0$ and $\varkappa, \omega$ in AREA III.

in this area, and hence P has a negative minimum at $z = -z_1$ and a negative minimum at $z = -z_2$. Except for these extrema of P, the characteristics of this mode are those of surface waves. It is to be presumed, on the basis of continuity, that the graph of Eq. (72–10) crosses from area I into area III and that the stability modifies the surface waves in this frequency range.

$n = -1$.

In this case, Q has a maximum in the range $-z_1 < z < z_2$. Since Q and dQ/dz must now have the same sign in order to satisfy the boundary condition at the surface, Q must have a node between this maximum and the surface. The figure shows this node between $z = -z_2$ and $z = -S$, but the reader will be able to construct an equally acceptable graph for which the node lies between the maximum and $z = -z_2$. The ratio of P and dQ/dz being negative, it follows that P has a negative minimum at $z = -z_1$, and a positive maximum at $z = -z_2$. It must also have a node at the maximum of Q; this being the only node of P, the ordinal number is $n = -1$.

The amplitude of the vertical motion being Q, the greatest motion will occur in the thermocline-channel. The amplitude will diminish rapidly to either side of the range $-z_1 < z < -z_2$; for the sake of clarity, this has not been indicated quantitatively on the graph. The motion can, perhaps, be described as an oscillation of the thermocline as a whole.

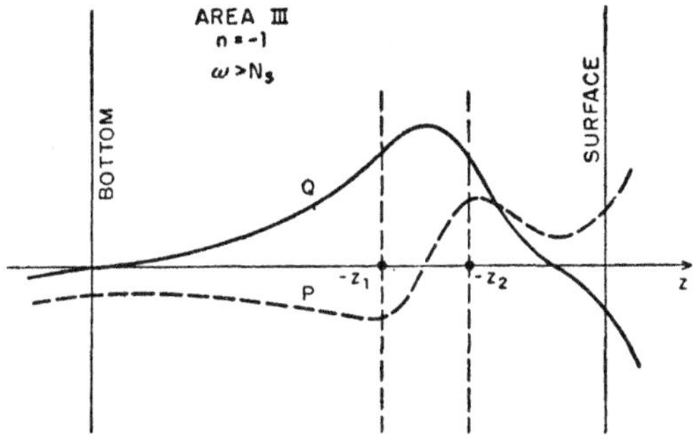

FIG. 46. Sketch of the eigensolution for $n = -1$ and $\varkappa,\omega$ in AREA III.

$n = -2$.

In this case Q has three nodes and P, two, which are in the range $-z_1 \leqslant z \leqslant -z_2$; this is shown on Fig. 47. The upper and lower parts of this range therefore oscillate 180° out of phase. The mode may be described as a pulsation of the thermocline layer which alternately expands and contracts.

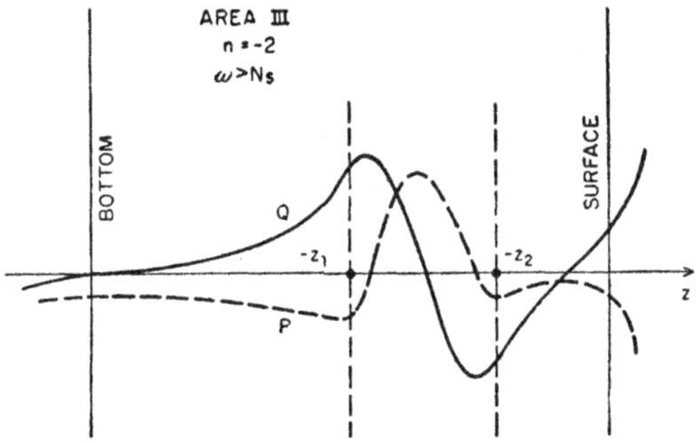

FIG. 47. Sketch of the eigensolution for $n = -2$ and $\varkappa,\omega$ in AREA III, $\omega > N_S$.

The larger negative ordinal numbers lead to modes of motion that are still more complex because of the greater number of nodes. The reader will be able to construct their diagrams according to the same principles used here.

Modification when the thermocline channel extends to surface or bottom. Thus far it has tacitly been assumed that both roots of $N^2 = \omega^2$ lie in the range $-S - D \leqslant z \leqslant -S$ that is actually occupied by water. Reference to Fig. 37 shows that, for sufficiently low values of ω, one or both of these roots can fall into the physically inadmissible ranges. For simplicity it will be supposed that the value of N at the surface, N_S, is greater than its value, N_B, at the bottom, as illustrated on Fig. 36. As ω decreases, $-z_2$ will increase and

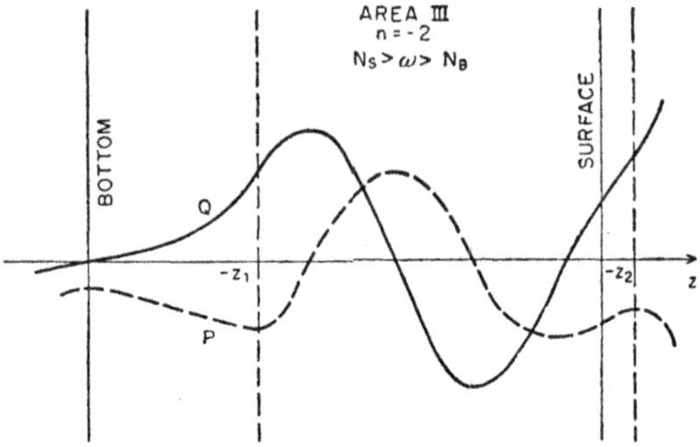

FIG. 48. Sketch of the eigensolution for $n = -2$ and $\varkappa,\omega$ in AREA III, $N_S > \omega > N_B$.

ultimately become greater than $-S$; a further decrease of ω will so decrease that $-z_1 < -S - D$. One will therefore have to consider the three cases

(1) $$\omega > N_S,$$

(2) $$N_S > \omega > N_B,$$

(3) $$N_B > \omega,$$

separately. Thus far only the case (1) has been treated.

No detailed treatment of the other two cases will be given since the reader should be able to supply them from what has been said above and an inspection of Figs. 48 and 49. These have been sketched for $n = -2$, $N_S > \omega > N_B$ and for $n = -2$, $N_B > \omega$, respectively.

AREAS IV AND V

It is also possible to construct qualitative diagrams for the eigen-solutions represented by points in the areas IV and V, but it is probable that quantitative considerations would show that for the

Earth's oceans no eigensolutions occur in these areas. In the case of area IV, this is because the Earth's oceans are shallow, in the sense of Section 55. It has already been seen that no waves can be represented by points in area V unless the deepest layers have sub-critical stability.

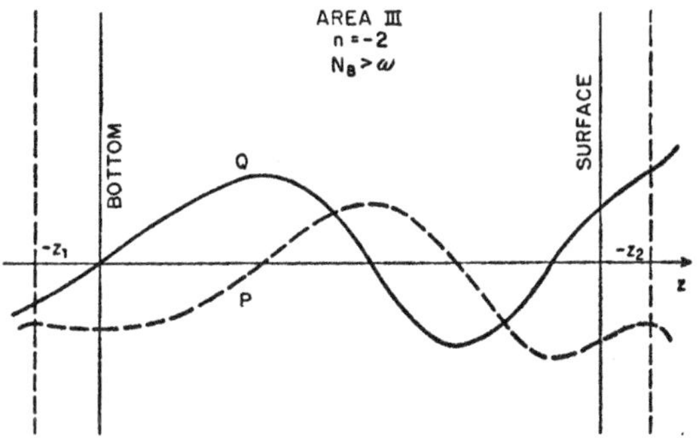

FIG. 49. Sketch of the eigensolution for $n = -2$ and $\varkappa, \omega$ in AREA III, $N_B > \omega$.

73. Analytic solution of the residual equations

It will now be explicitly supposed that the stability parameter, N, is given by Eq. (67–1):

$$N^2 = -N_M{}^2[(2Z/z) + (Z/z)^2]. \tag{1}$$

Then Eq. (72–4) becomes

$$\frac{d^2Q}{dz^2} = -\frac{1}{c^2}\left(1 - \frac{\varkappa^2c^2}{\omega^2 - \Omega^2}\right)[\omega^2 + N_M{}^2(2Z/z) + N_M{}^2(Z/z)^2]\,Q. \tag{2}$$

The analytic solution of this equation can be expressed in terms of Whittaker's functions; these satisfy the differential equation

$$\frac{d^2W}{d\xi^2} + \left\{-\frac{1}{4} + \frac{k}{\xi} + (\tfrac{1}{4} - l^2)\frac{1}{\xi^2}\right\}W = 0 \tag{3}$$

(Ref. 7, Chapter 16, 23, p. 88). Equation (73–2) reduces to this if the variables are defined as follows:

$$\begin{aligned} \xi &= -z/\alpha, \\ \alpha &= (c/2\omega)[(\omega^2 - \Omega^2)/(\varkappa^2c^2 + \Omega^2 - \omega^2)]^{\frac{1}{2}}, \\ k &= N_M{}^2Z/2\omega^2\alpha, \\ l^2 - \tfrac{1}{4} &= N_M{}^2Z^2/4\omega^2\alpha^2 = \tfrac{1}{2}kZ/\alpha. \end{aligned} \tag{4}$$

Two independent special solutions of Eq. (73–3) are the Whittaker functions $W(k, l, \xi)$ and $W(-k, -l, -\xi)$; a somewhat more detailed discussion of these functions will be given in Section 79. The most general solution of Eq. (73–2) is therefore

$$Q = AW(k, l, -z/\alpha) + BW(-k, l, z/\alpha), \tag{5}$$

A and B being constants. It should be noted that α, k and l may be either real or pure imaginary numbers.

The second of Eqs. (72–1) then yields the expression for P:

$$P = [4\omega^2\alpha/c][AW'(k, l, -z/\alpha) - BW'(-k, l, z/\alpha)], \tag{6}$$

where the notation

$$W'(k, l, \xi) = \partial W(k, l, \xi)/\partial \xi$$

has been used.

It remains to impose the boundary conditions, Eqs. (72–2, 3):

$$Pc + gQ = 0 \quad \text{at} \quad z = -S,$$

$$Q = 0 \quad \text{at} \quad z = -S - D.$$

Substitution from Eqs. (73–5, 6) reduces these to

$$A\{W(k, l, S/\alpha) + (4\omega^2\alpha/g)W'(k, l, S/\alpha\}$$
$$+ B\{W(-k, l, -S/\alpha) - (4\omega^2\alpha/g)W'(-k, l, -S/\alpha)\} = 0. \tag{7}$$

$$AW[k, l, (S + D)/\alpha] + BW[-k, l, -(S + D)/\alpha] = 0. \tag{8}$$

In order that these equations shall have a non-trivial solution for A and B, their determinant must vanish:

$$W[k, l, (S + D)/\alpha]\{W(-k, l, -S/\alpha) - (4\omega^2\alpha/g)W^1(-k, l, -S/\alpha)\}$$
$$-W[-k, l, -(S + D)/\alpha]\{W(k, l, S/\alpha) + (4\omega^2\alpha/g)W'(k, l, S/\alpha)\} = 0. \tag{9}$$

This equation results from the two boundary conditions at the surface and bottom. It is therefore analogous to the equations studied in Chapter X and to Eqs. (72–10, 12) encountered above. Unfortunately, the theory of the Whittaker functions is not sufficiently advanced to make possible a graphic treatment of Eq. (73–9) that is analogous to those of Figs. 26, 28 and 30. Consequently, an approximate calculation, based on the W-K-B approximation, must be substituted; this follows in the next section.

74. Further application of the W-K-B approximation

The approximate calculation, mentioned at the end of the last section is based on a very simple idea. In those regions of the z-axis where the graph of Q is oscillatory, the W-K-B approximation is

$$Q = Q_0 \sin (S + \text{const.}), \tag{1}$$

if x, y and t are not varied. Hence, if S increases by π units in a certain range of the z-axis, Q will have exactly one root or node in that range. Therefore

$$[S(z'') - S(z')]/\pi \tag{2}$$

is an approximation to the number of nodes in the interval $z' \to z''$.

Recalling that

$$S(z) = \int \gamma \, dz \tag{3}$$

and letting m be the number of nodes in $z' - z''$, one may write

$$\int_{z'}^{z''} \gamma \, dz = (m + \Delta)\pi \tag{4}$$

where Δ is the error in the various approximations involved in Eqs. (74–1) and (74–2), and may again be called the ordinal defect. There are refinements of the W-K-B method that enable the calculation of Δ in certain cases.[5]

The acoustic modes of area II. Equation (74–4) will first be applied to those eigensolutions that represent acoustic waves. It has been seen that these occur for $(\varkappa,\omega)$ points in area II and that the nth mode has exactly n nodes in the range $-S - D \leqslant z \leqslant -S$. Using this observation and Eq. (68–2) for γ, Eq. (74–4) becomes

$$\frac{1}{c}\left\{1 - \frac{\varkappa^2 c^2}{\omega^2 - \Omega^2}\right\}^{\frac{1}{2}} \int_{-S-D}^{-S} (\omega^2 - N^2)^{\frac{1}{2}} \, dz = (n + \Delta)\pi. \tag{5}$$

As was seen above, the integral on the left of Eq. (74–4) can be evaluated rigorously in terms of elementary functions. However, if $\omega \gg N$ as is usually the case, the integrand is approximately equal to ω with a proportional error that is of the order of magnitude $(N/\omega)^2$. To this approximation, Eq. (74–5) becomes

$$\frac{\omega D}{c}\left\{1 - \frac{\varkappa^2 c^2}{\omega^2 - \Omega^2}\right\}^{\frac{1}{2}} = (n + \Delta)\pi.$$

Under these conditions ω is also much greater than Ω so that this is essentially

$$\left\{\frac{\omega^2}{c^2} - \varkappa^2\right\}^{\frac{1}{2}} = (n + \Delta)\pi/D. \qquad (6)$$

If $\Delta = -\frac{1}{2}$, this is the "open organ pipe" formula already discussed in Section 55.

This result is instructive for it shows that the ordinal defect need not always be small compared to unity. For small ordinal numbers, it therefore becomes important to devise methods for calculating Δ. For large ordinal numbers, the approximation $\Delta = 0$ will still cause only a small proportionate error.

Because of the shallowness of the Earth's oceans, it is unnecessary to consider the more elaborate formulae that result from the rigorous evaluation of the integral.

The internal gravity modes of area III. In Area III, $\omega < N_M$, and hence Eq. (74–5) must be replaced by

$$\frac{1}{c}\left\{\frac{\varkappa^2 c^2}{\omega^2 - \Omega^2} - 1\right\}^{\frac{1}{2}} \int_{z'}^{z''} (N^2 - \omega^2)^{\frac{1}{2}}\, dz = (m + \Delta)\pi \qquad (7)$$

and the integral must be evaluated without further approximation. Moreover, the definitions of z', z'' and m will be different for the three cases illustrated by Figs. 47, 48 and 49.

<u>$\omega > N_S$</u>

In this case $z' = -z_1$, $z'' = -z_2$ and $m = |n| - 1$, as follows from inspection of Fig. 37. Using the notation of Section 70 Eq. (74–7) becomes

$$-\frac{N_M Z}{c}\left\{\frac{\varkappa^2 c^2}{\omega^2 - \Omega^2} - 1\right\}\int_{1-b}^{1+b} \frac{(a^2 - 2u + u^2)^{\frac{1}{2}}}{u^2}\, du$$
$$= (|n| - 1 + \Delta)\pi. \qquad (8)$$

Using Eqs. (70–6, 7) the integral is readily evaluated and Eq. (74–8) becomes

$$\frac{N_M Z}{c}\frac{N_M - \omega}{\omega} = (|n| - 1 + \Delta)\left\{\frac{\varkappa^2 c^2}{\omega^2 - \Omega^2} - 1\right\}^{-\frac{1}{2}}. \tag{9}$$

<u>$N_S > \omega > N_B$</u>

In this case inspection of Fig. 37 shows that $z' = -z_1$, $z'' = -S$, $m = |n|$. Hence, in Eq. (74–6) the upper limit of the integral must be replaced by

$$u_S = Z/S, \tag{10}$$

and the right side by $(|n| + \Delta)\pi$. Defining ϕ_S and ψ_S by

$$Z/S = 1 + b \sin \phi_S = a^2/(1 + b \sin \psi_S), \tag{11}$$

the result is that (cf. Eq. 70–7)

$$\frac{N_M Z}{c}\left\{\frac{\pi}{2}\frac{N_M - \omega}{\omega} - \phi_S - (\psi_S + b \cos \psi_S)/a\right\}$$
$$= (|n| + \Delta)\pi \left\{\frac{\varkappa^2 c^2}{\omega^2 - \Omega^2} - 1\right\}^{-\frac{1}{2}}. \tag{12}$$

<u>$N_B > \omega$</u>

In this case, $z' = -S - D$, $z'' = -S$, $m = |n| + 1$; with the definitions

$$Z/(S + D) = 1 + b \sin \phi_B = a^2/(1 + b \sin \psi_B), \tag{13}$$

$$\frac{N_M Z}{c}\{\phi_B + (\psi_B + b \cos \psi_B)/a - \phi_S - (\psi_S + b \cos \psi_S)/a\}$$
$$= (|n| + 1 + \Delta)\pi \left\{\frac{\varkappa^2 c^2}{\omega^2 - \Omega^2} - 1\right\}^{-\frac{1}{2}}. \tag{14}$$

These results are illustrated numerically in Fig. 51. This has been prepared with the numerical values

$$2\pi/N_M = 15 \text{ min},$$
$$2\pi/N_S = 30 \text{ min},$$
$$2\pi/N_B = 60 \text{ min},$$
$$2\pi/\Omega = 12 \text{ hr},$$
$$D = 4 \text{ km}.$$

When these data are combined with Eq. (67–1), it is found that

$$Z = 133{\cdot}5 \text{ m},$$

$$S = 71{\cdot}5 \text{ m},$$

$$Z - S = 62 \text{ m}.$$

By comparison with the Earth's oceans this thermocline is very shallow, and the stability at the bottom is very great. It would be desirable to consider more realistic numerical examples, but this would require computations that have not been made.[1]

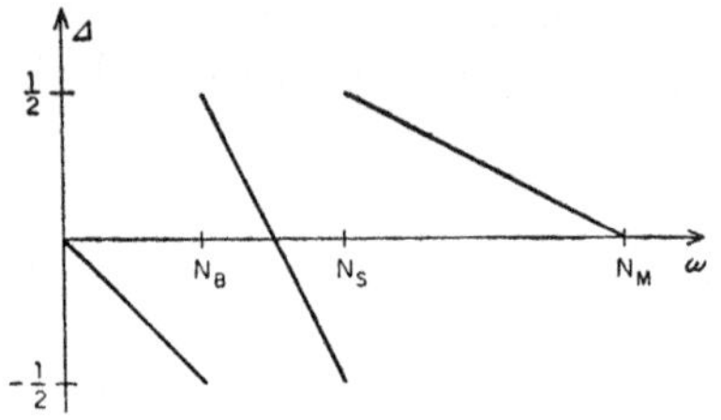

FIG. 50. The assumed variation of the ordinal defect, Δ, with frequency, ω.

In order to obtain reasonable results for the smaller ordinal numbers, it is also necessary to make some estimate of the ordinal defects in Eqs. (74–9, 12 and 14). Again, this should be the result of computations that have not been made; consequently it will arbitrarily be assumed that Δ is independent of n and depends on ω as shown by Fig. 50. The discontinuities of this graph have been chosen so that Eqs. (74–8 and 12) yield the same value of $\varkappa$ for $\omega = N_S$ and Eqs. (74–12 and 14) the same value for $\omega = N_B$.

Figure 51 shows the graphs of ω as a function of $\varkappa$ and n obtained in this way. The effect of the arbitrary assumption concerning the ordinal defect is seen in the discontinuities of the slopes of the curves. The curve for $n = -1$ is appreciably in error because of this arbitrariness, but the proportionate error diminishes with increasing $|n|$.

This figure is to be compared with Fig. 32; the same logarithmic scale has been used in the two, but the origins are displaced by one decade along the wave-length axis. The difference between the case of constant N, and the thermocline case is seen to be slight for periods near 12 hr. As the period decreases, the differences become greater.

For given wave-length and ordinal number, Fig. 51 shows much longer periods than does Fig. 32.

It should be noticed that Eq. (58–3), on which Fig. 32 is based, could have been derived by the methods of this section, using $N =$ constant, $\Delta = 0$, $z'' = -S$, $z' = -S - D$ in the integral of Eq. (74–7).

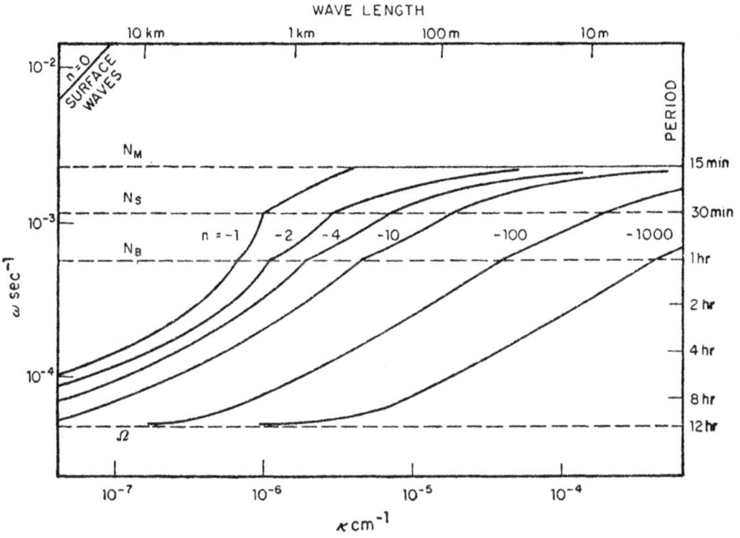

FIG. 51. Portion of the diagnostic diagram for an ocean with thermocline and constant depth, $D = 4$ km. (W-K-B approximation.)

75. The two-layer model

It has been noted that Eq. (54–17), for the variation of N with depth, results in somewhat unrealistic values. The thermocline is too shallow and too thick, and the stability of the deep layers is too great, when compared with the values of Fig. 10.

One may go to the other extreme and consider the case of an extremely high and narrow thermocline. This, because of Eq. (24–13) results in a very large gradient of density at the thermocline, which may be approximated by a discontinuity of density. Thus one arrives at a two-layer model of the thermocline, consisting of two fluids of differing densities, the lighter floating on the heavier. This model has been discussed (for $\Omega = 0$) in detail by Lamb (Ref. 4, pp. 370–375) and others. Ufford[6] has made empirical measurements on internal waves and interpreted them in terms of this model.

It is found that this model yields only two eigensolutions, for $n = 0$ and $n = -1$ in the present notation. More precisely, the eigensolutions for $n \leqslant -2$ all have the frequency $\omega = 0$, for all values of $\varkappa$. This is the ultimate extreme of the period increase due to the thermocline.

Thermoclines also occur in the atmosphere, although they are less permanent than is the oceanic thermocline. In meteorology, they are often called temperature inversions; the stratosphere, Fig. 14, *A*, may also be considered as a thermocline and would then be the best analogue of the permanent thermocline of the ocean. However, the discussion of the stratosphere is complicated by factors that do not occur in the ocean and will be postponed.

The temperature inversions are the site of internal waves, as follows at once from the foregoing discussion. These waves are often made visible by cloud formations—the so-called mackerel sky. In certain regions, the temperature inversion is present for longer intervals, and other phenomena may be observed. Gossard and Munk(3) have discussed some of these and shown that the associated clouds move with the velocity of the internal waves, not with the velocity of the wind. Qualitatively, similar cloud formations are often observed in other parts of the world, but have not been investigated quantitatively.

References for Chapter XII

1. J. E. Fjeldstad, Interne Wellen, *Geofysiske Publikasjoner* **10,** No. 6 (1933) (Oslo).
2. P. Groen, Contribution to the theory of internal waves, *Koninklijk, Nederlands, Meteorologisch Institut de Bilt, Medelingen en Vorhandelingen* B, No. 225.
3. E. Gossard and W. H. Munk, On gravity waves in the atmosphere, *J. Meteorology* **11,** 259 (1954).
4. H. Lamb, *Hydrodynamics*, Cambridge Univ. Press, 6th ed., 728 pp. 1932.
5. P. M. Morse and H. Feschbach, *Methods of Theoretical Physiks*, pp. 528, 583, 870–880 McGraw-Hill, New York, 1953.
6. C. W. Ufford, Internal Waves in the Ocean, *Trans. Am. Geophys. Union* **28,** pp. 79-99 (1947).
7. E. T. Whittaker and G. N. Watson, *A Course of Modern Analysis*, Cambridge Univ. Press, 1927.

CHAPTER XIII

THE THERMOSPHERE

76. Introduction

THE preceding chapter treated the effect of the most outstanding oceanic feature—the thermocline—on the eigensolutions. The stratification of the atmosphere exhibits so many features that it is difficult to isolate any one of them for a similar treatment. However, there is one feature that, by itself, causes the eigensolutions of the Earth's atmosphere to differ from those of the isothermal atmosphere treated in Chapters VIII and IX.

This is the thermosphere—the high altitude region where the temperature increases, seemingly without limit. The theory of atmospheres falls naturally into two parts. One of these treats those atmospheres whose temperature becomes asymptotically constant as the altitude increases. The other treats atmospheres that have a thermosphere whose temperature increases without limit as the altitude increases. In the first case, the simplest possible example is the isothermal atmosphere already treated above. The simplest example for the other case is an atmosphere with a constant, positive temperature gradient—an atmosphere that has no troposphere, no stratosphere, but only a thermosphere. This chapter will be devoted to the discussion of such a "pure" thermosphere.

The difference between a thermosphere and an isothermal atmosphere can most simply be stated by reference to the diagnostic $(\varkappa,\omega)$ diagram. It has already been seen that in the isothermal atmosphere, all points of certain areas of this diagram correspond to eigensolutions (Sections 42 and 51). This is not true of an ocean. In the oceanic case, the second boundary condition at the free surface modified the $(\varkappa,\omega)$ diagram, so that only points on the modal curves correspond to eigensolutions (Chapters X and XII).

The existence of a thermosphere also brings a second boundary condition with it—the "natural" boundary condition at infinity. The thermosphere is, in a sense, a "lid" on the earth's atmosphere, in that it reflects energy radiated upward and prevents it from escaping to infinity. In the isothermal atmosphere, the rays are straight lines, extending to infinity. Energy moving upward along one of them will eventually escape. In the thermosphere, on the other hand, the rays are generally curved, and except for a very few, run their entire course between finite limiting altitudes (see Fig. 52). They do not

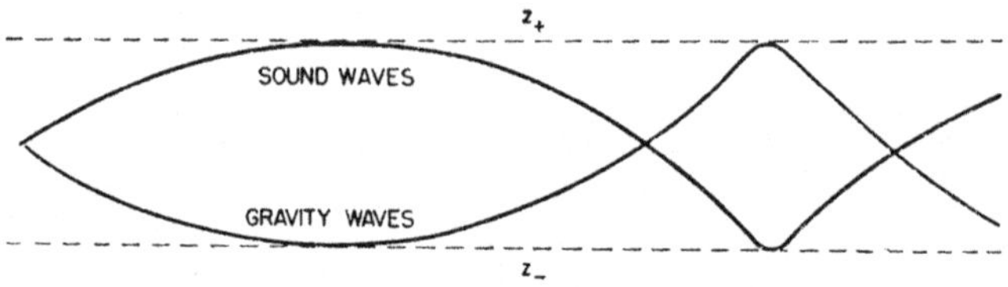

FIG. 52. Two rays in the thermosphere.

extend to infinite altitudes, and therefore energy moving along one of them cannot escape, but is trapped in the lower layers of the atmosphere. This is a fundamental physical difference between atmospheres with and without a thermosphere.

This makes itself apparent in the nature of the eigensolutions. In the case of any atmosphere with a thermosphere, these are limited (Sections 32 and 34). They are modes of motion that have a finite energy per unit area. In the case of spherical level surfaces, they even have a finite total energy. In the case of an atmosphere without thermosphere, most eigensolutions represent modes of motion that have infinite energies.

This chapter will be devoted to a demonstration of these facts. The first sections will consider the rays and their reflection by the temperature gradient. Then the "natural boundary condition" will be derived rigorously. This will be found too complicated for numerical discussion, and the modal curves will therefore be derived by the W-K-B approximation, as in Section 74.

This will be done only for the constant positive temperature gradient and plane level surfaces, more general cases being postponed to Chapter XV. As in Section 8, let the zero-order temperature be given by

$$R\theta_0 = c^2/\gamma = gz/\nu, \quad \nu = \text{const.}, > 0, \tag{1}$$

so that the rigid lower boundary of the atmosphere is at $z = S > 0$. (Here γ is the ratio of the specific heats.) As in the previous chapter, all values of z will be considered, but only those greater than S will be of physical significance.

The formulae of Section 25, for the parameters N and Γ, may be written

$$N^2/g = [1 + \nu(\gamma - 1)/\gamma]/z = \gamma r^2/z\nu, \tag{2}$$

$$\Gamma = [\nu(2 - \gamma)/\gamma - \tfrac{1}{4}]/z = s/z. \tag{3}$$

Here γ is the ratio of the specific heats, and the parameters ν, r and s are dimensionless constants. Another convenient parameter is

$$m = \gamma r/\nu. \tag{4}$$

If a temperature gradient of 10 deg/km is postulated,

$$\nu = 3{\cdot}42, \quad s = 1{\cdot}22, \quad r = 2{\cdot}20, \quad m = 0{\cdot}90,$$

all having the general order of magnitude of unity.

77. The case of no rotation

The investigation will begin with the theory of simple waves, and these will be studied with the aid of ray theory. The formal calculation of the rays is relatively simple, for the integral that must be evaluated is an elementary one. It will be found that the rays are prolate cycloids, curves that have a simple geometric construction. They can be generated by a circle that rolls on a line: any point inside the rolling circle traces a prolate cycloid. The problem is thus reduced to finding the radius of the circle, the line traced by its centre, and the distance of the tracing point from the centre. These will all be functions of the frequency, ω, and the horizontal wave number $\varkappa$.

Less simple than the formal calculation is the interpretation of the formulae. In particular, it becomes necessary to distinguish between those rays that intersect the ground, and those that do not. Those that do intersect the ground are, of course, reflected there.

Formal calculation of the rays. Equation (63–4), for $\Omega = 0$, is

$$\gamma^2 c^2 = (\omega^2 - N^2 - \Gamma^2 c^2) + \varkappa^2 c^2 (N^2 - \omega^2)/\omega^2, \tag{1}$$

γ here and in the following again being the vertical component of the propagation vector. Defining

$$u = \log\,(\omega^2/mg\varkappa) \tag{2}$$

and using the formulae of the last section, the first equation may be written

$$\gamma^2 = (-1/z^2)(\varkappa^2 z^2 - 2r\varkappa z \cosh u + r^2 + s^2). \tag{3}$$

The roots of γ define the altitudes at which the rays are horizontal, and turn from ascent to descent or conversely. They are given by

$$\varkappa z_{\pm} = r \cosh u \pm a \tag{4}$$

where

$$a^2 = r^2 \sinh^2 u - s^2.$$

To evaluate the integrals required in the following, it is convenient to transform from z to the variable θ, defined by

$$\varkappa z = r \cosh u + a \sin \theta, \tag{5}$$

so that Eq. (77–3) finally reduces to

$$\gamma z = a \cos \theta. \tag{6}$$

The equation of the rays is (cf. Eq. (64–9))

$$\begin{aligned} \varkappa(x - x_0) &= (\varkappa^2/\omega^2) \int \frac{\omega^2 - N^2}{\gamma} \mathrm{d}z \\ &= \int \frac{\varkappa z - r \exp(-u)}{\gamma z} \mathrm{d}(\varkappa z). \end{aligned} \tag{7}$$

Using Eqs. (77–5) and (77–6), this integral is readily evaluated:

$$\varkappa(x - x_0) = r\theta \sinh u - a \cos \theta. \tag{8}$$

Equations (77–5) and (77–8) are the equations of a prolate cycloid. The radius of the rolling circle is $b = r \sinh u$ and its centre traces the line $\varkappa z = r \cosh u$. The point that traces the ray is at a distance a from the centre of the circle, and is in the interior of this circle. Two cases must be distinguished; if $u > 0$, the circle rolls on a horizontal line that is below its centre, and it will be seen that this corresponds to sound waves. If $u < 0$, the circle rolls on a line that is above its centre, and these are the rays of the gravity waves. A typical example of each kind is shown on Fig. 52. It will be noted that the radius of the circle, and the elevation of its centre are independent of the sign of u; consequently sound rays and gravity rays are associated in pairs that have the same value of $|u|$. It is such a pair that is shown on Fig. 52.

Interpretation of the results. To correlate these results with those obtained in earlier chapters, one will wish to construct a $(\varkappa, \omega)$ diagram. As a preliminary step, a study of Eq. (77–4) is indicated, since γ is real only in the range $z_- \rightarrow z_+$. Figure 53 shows a graph of Eq. (77–4); it consists of two branches, each of which is asymptotic to one of the curves

$$\varkappa z = r e^{\pm u},$$

as shown. For each value of $|u| > u_c$, there are two points on the curve; the larger ordinate is $\varkappa z_+$, the smaller, $\varkappa z_-$. The value of u_c is given by

$$\sinh u_c = s/r. \tag{9}$$

When $|u| < u_c$, the values of $\varkappa z_\pm$ are complex, and there are no values of z for which γ is real.

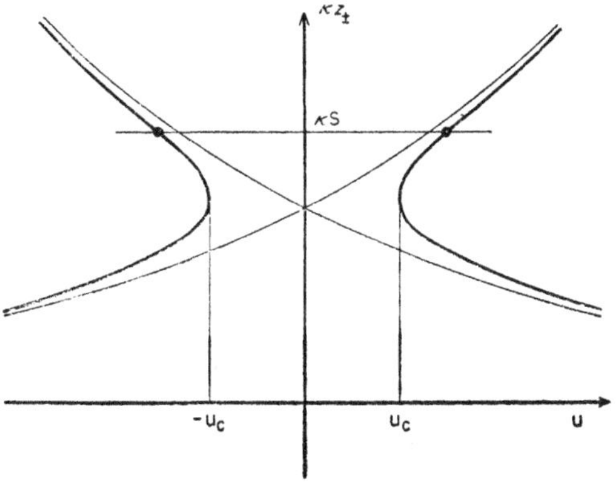

FIG. 53. Graph of the equation $\varkappa z_\pm = r \cosh u \pm a$, where $a^2 = r^2 \sinh^2 u - s^2$.

These results can be transferred to the $(\varkappa,\omega)$ diagram; from Eq. (77–2), it follows that $u = \pm u_c$ is equivalent to

$$\omega^2 = \varkappa m g e^{\pm u_c} \tag{10}$$

and this is the equation of two parabolae, as shown on Fig. 54. Consequently, there are no rays that correspond to points $(\varkappa,\omega)$ that lie between these two parabolae.

Formally, there is a ray for every point above the parabola $u = u_c$, and for every point below $u = -u_c$. However, not all of these rays are physically realizable, since they may run their whole course below $z_+ = S$, and hence lie outside the volume occupied by the fluid.

Moreover, those rays for which $z_- < S$ and $z_+ > S$ are reflected at the ground. To examine these matters, one may return to the graph of Fig. 53 and consider the two points whose ordinates are $\varkappa S$. If

$$\varkappa S \geqslant r \cosh u + a = \varkappa z_+,$$

then $z_+ \leqslant S$ and the ray is not physically possible. If

$$\varkappa S \leqslant r \cosh u - a = \varkappa z_-$$

then $z_- \geqslant S$, the entire ray only extends above ground, and reflection does not occur. One will therefore transfer the equation

$$\varkappa S = r \cosh u \pm a = \varkappa z_\pm \tag{11}$$

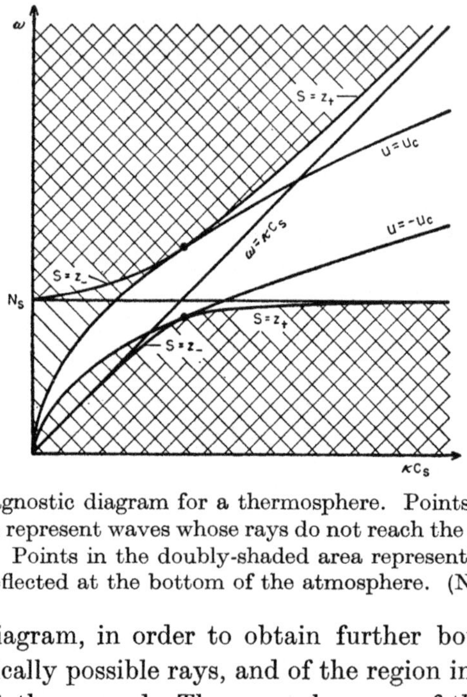

FIG. 54. Diagnostic diagram for a thermosphere. Points in the singly-shaded areas represent waves whose rays do not reach the bottom of the atmosphere. Points in the doubly-shaded area represent waves whose rays are reflected at the bottom of the atmosphere. (No rotation.)

to the $(\varkappa,\omega)$ diagram, in order to obtain further boundaries of the region of physically possible rays, and of the region in which the rays are reflected at the ground. The general course of the graph of this equation is also shown on Fig. 54.

On this diagram, the value of the velocity of sound at the ground level is given by

$$c_S = (\gamma g S/\nu)^{\frac{1}{2}}.$$

Similarly,

$$N_S{}^2 = \gamma r^2/\nu S$$

and N_S is the value of N at ground level. Both N_S and c_S are extreme values for this atmosphere. It is seen that the graph of $S = z_+$ has the lines $\omega = \varkappa c_S$ and $\omega = N_S$ as asymptotes. The graph of $S = z_-$ intersects the ω-axis at $\omega = N_S$ and $\omega = 0$. Their tangents at these intersections have the slopes 0 and c_S, respectively.

Points of this $(\varkappa,\omega)$ diagram that lie in the unshaded area do not correspond to physically possible rays. Points in the doubly-shaded area correspond to rays that are reflected at the ground level, $z = S$. Points in the singly-shaded area correspond to rays that run their entire course aloft, and are thus not reflected at the ground.

Comparison with the $(\varkappa,\omega)$ diagrams of earlier chapters will make it evident, without further detailed calculations, that the upper area, where $u > u_c$, corresponds to acoustic waves, and the lower area, where $u < -u_c$, to gravity waves. In earlier chapters, it has also been noted that no rays correspond to the Lamb waves; it will therefore be anticipated that those values of $(\varkappa,\omega)$ that correspond to these eigensolutions will lie on a curve in the unshaded area of Fig. 54.

Returning to Fig. 52, which shows examples of the rays, it is to be noted that both rays are confined to the same channel, and have horizontal tangents at its boundaries $(z = z_\pm)$. One vertex of each ray has a small radius of curvature, but neither vertex is a cusp, such as is shown on Fig. 40 for the idealized ocean. The reason for this is found in the fact that $\Gamma \neq 0$ and $c \neq$ constant in this atmospheric case. For large values of $|u|$, $\Gamma \to 0$ while a and $r \sinh u$ become nearly equal. For practical purposes, the sharper vertex then becomes a cusp. The rays then become simple cycloids—a result that is well known in the ordinary theory of atmospheric acoustics. The reason why the other vertex retains a finite radius of curvature is to be found in the variability of the velocity of sound, which contributes to the refraction of the rays, and affects the positions of the channel boundaries.

The most striking and important difference between these results and those for the isothermal atmosphere has been noted in the introduction. In the isothermal atmosphere, the rays are straight lines; unless they happen to be horizontal, they extend to infinite altitudes. Since they are the streamlines of the energy flow, energy is radiated to or from infinity. In the present case, the rays are curved, and do not reach infinite altitudes. Energy is therefore not radiated to infinity.

Thus far, no evidence for the "natural boundary condition" has been presented; this will be given in Sections 79 and 80. In Sections 86 and 87 there will be an additional discussion of the eigensolutions. In the following section, the concern will be about the manner in which rotation about a vertical axis modifies the rays.

78. Vertical axis of rotation

If $\Omega \neq 0$, and the axis is vertical, the expression corresponding to Eq. (77–1) is

$$\gamma^2 c^2 = \omega^2 - N^2 - \Gamma^2 c^2 + b^2 \varkappa^2 c^2 (N^2 - \omega^2)/\omega^2 \tag{1}$$

where the abbreviation

$$b^2 = \omega^2/(\omega^2 - \Omega^2) \tag{2}$$

has been used. It reduces to

$$\gamma^2 = (-1/z^2)(b^2\varkappa^2 z^2 - 2rb\varkappa z \cosh u + r^2 + s^2), \tag{3}$$

provided u is now defined by

$$u = \log (\omega^2/mg\varkappa b). \tag{4}$$

The channel boundaries are given by

$$b\varkappa z_{\pm} = r \cosh u \pm a, \tag{5}$$

a having the same definition as before. Since $b > 1$, the separation $z_+ - z_-$ is less than if there were no rotation.

The equation of the rays is

$$\varkappa(x - x_0) = (\varkappa^2 b^2/\omega^2) \int \frac{\omega^2 - N^2}{\gamma} \, dz, \tag{6}$$

or, if

$$b\varkappa z = r \cosh u + a \sin \theta, \tag{7}$$

$$\gamma z = a \cos \theta, \tag{8}$$

then

$$\varkappa(x - x_0) = r\theta \cosh u - a \cos \theta. \tag{9}$$

The rays are therefore flattened prolate cycloids, the flattening having the ratio $1 : b$. Sound waves, for which b is very nearly unity, are thus not appreciably affected by the rotation, nor are gravity waves for which $\omega^2/\Omega^2 \gg 1$. But if ω^2 approaches Ω^2, the rays become less inclined to the horizontal, and the channel is displaced toward smaller values of z.

The interpretation of these results and the construction of the $(\varkappa, \omega)$ diagram proceeds in a very similar way to that of the previous section. The only difference is that the parabolae are replaced by the curves

$$\omega(\omega^2 - \Omega^2)^{\frac{1}{2}} = mg\varkappa e^{\pm u_c}. \tag{10}$$

The general character of the resulting diagram is shown on Fig. 55.

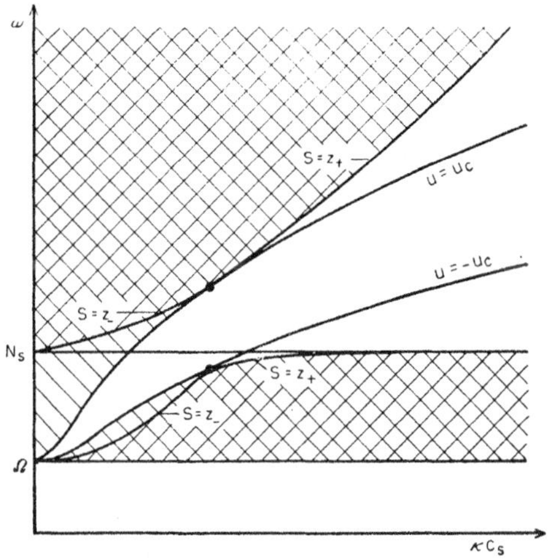

FIG. 55. Diagnostic diagram for a thermosphere with rotation. Legend as for Fig. 54.

Thus far, the tacit assumption has been that $\omega^2 > \Omega^2$. If $\omega^2 < \Omega^2$, the definitions must be altered:

$$b^2 = \omega^2/(\Omega^2 - \omega^2), \tag{11}$$

$$\omega(\Omega^2 - \omega^2)^{\frac{1}{2}} = g\varkappa m e^u. \tag{12}$$

Since the left-hand side of Eq. (77–11) is small, u can only be negative. These definitions result in

$$\gamma^2 z^2 = (b\varkappa z - r \sinh u)^2 - (r^2 \cosh^2 u + s^2) \tag{13}$$

so that the channel boundaries are given by

$$b\varkappa z_{\pm} = -r \sinh u \pm a, \tag{14}$$

$$a^2 = r^2 \cosh^2 u + s^2. \tag{15}$$

It will be verified that $z_- < 0$ and therefore without physical significance. The channel consists of those values of z greater than z_+, which has a large positive value:

$$z_+ \geqslant 2\, mrg/\Omega^2 = mr \times 10^6 \text{ km}.$$

The rays will come from infinity, turn at z_+ and recede to infinity. They run their entire course in the regions of sub-critical stability, $N < \Omega$. It may well be doubted that such rays have any significance for the Earth's atmosphere, but, from a formal standpoint, they do constitute exceptions to the statements made in the introduction.

79. Solution of the residual equations

The residual equations,

$$\begin{aligned} c\left(\frac{\mathrm{d}}{\mathrm{d}z} + \Gamma\right)P &= (N^2 - \omega^2)Q, \\ c\left(\frac{\mathrm{d}}{\mathrm{d}z} - \Gamma\right)Q &= [1 - \varkappa^2 c^2/(\omega^2 - \Omega^2)]P, \end{aligned} \tag{1}$$

can be solved rigorously if N and Γ have the form of Eqs. (76–2, 3). The general ideas underlying the following calculation are explained in Ref. 2, Sections 49, 50.

Retaining the definitions of the previous sections, the Eqs. (79–1) are

$$\begin{aligned} \left(\frac{\mathrm{d}}{\mathrm{d}z} + \frac{s}{z}\right)P &= \left(\frac{g\gamma r}{\nu z} - \omega^2\right)\left(\frac{\nu}{g\gamma z}\right)^{\frac{1}{2}}Q, \\ \left(\frac{\mathrm{d}}{\mathrm{d}z} - \frac{s}{z}\right)Q &= \left(1 - \frac{g\gamma z}{\nu\alpha}\right)\left(\frac{\nu}{g\gamma z}\right)^{\frac{1}{2}}P. \end{aligned} \tag{2}$$

The change of variables,

$$Q = Q_1 N^{-\frac{1}{2}}, \quad P = P_1 N^{+\frac{1}{2}}, \tag{3}$$

transforms Eq. (79–2) into

$$\begin{aligned} \left(\frac{\mathrm{d}}{\mathrm{d}z} + \frac{s - \frac{1}{4}}{z}\right)P_1 &= \left(\frac{r}{z} - \frac{\omega^2}{gm}\right)Q_1, \\ \left(\frac{\mathrm{d}}{\mathrm{d}z} - \frac{s - \frac{1}{4}}{z}\right)Q_1 &= \left(\frac{r}{z} - \frac{mg}{\alpha^2}\right)P_1, \end{aligned} \tag{4}$$

where

$$\alpha^2 = (\omega^2 - \Omega^2)/\varkappa^2. \tag{5}$$

It follows from the definitions given in the introduction that

$$(s - \tfrac{1}{4})^2 + r^2 = \tfrac{1}{4}(\nu + 1)^2. \tag{6}$$

This suggests the definition of a new parameter, ψ, such that

$$\begin{aligned} s - \tfrac{1}{4} &= \tfrac{1}{2}(\nu + 1) \cos \psi, \\ r &= \tfrac{1}{2}(\nu + 1) \sin \psi. \end{aligned} \tag{7}$$

If the dimensionless variable

$$\xi = 2\omega z/\alpha \tag{8}$$

is introduced, the Eqs. (79–4) are converted into

$$\begin{aligned} 2\frac{\mathrm{d}P_1}{\mathrm{d}\xi} + \frac{\nu + 1}{\xi}(P_1 \cos \psi - Q_1 \sin \psi) &= -Q_1 \mathrm{e}^u, \\ 2\frac{\mathrm{d}Q_1}{\mathrm{d}\xi} - \frac{\nu + 1}{\xi}(P_1 \sin \psi + Q_1 \cos \psi) &= -P_1 \mathrm{e}^{-u}, \end{aligned} \tag{9}$$

where u is defined by Eq. (78–4). It will be noted that ξ and u are real when $\omega^2 > \Omega^2$, and complex when $\omega^2 < \Omega^2$. The case $\omega^2 = \Omega^2$ is singular and requires separate investigation.

A second change of dependent variables:

$$\begin{aligned} P_1 &= f \cos \tfrac{1}{2}\psi + g \sin \tfrac{1}{2}\psi, \\ Q_1 &= -f \sin \tfrac{1}{2}\psi + g \cos \tfrac{1}{2}\psi, \end{aligned} \tag{10}$$

reduces the Eqs. (79–9) to the form

$$\begin{aligned} 2\frac{\mathrm{d}f}{\mathrm{d}\xi} + \frac{\nu + 1}{\xi} f &= af - b_1 g, \\ 2\frac{\mathrm{d}g}{\mathrm{d}\xi} - \frac{\nu + 1}{\xi} g &= -b_2 f - ag, \end{aligned} \tag{11}$$

where

$$\begin{aligned} a &= \cosh u \sin \psi, \\ b_1 &= \cosh u \cos \psi + \sinh u, \\ b_2 &= \cosh u \cos \psi - \sinh u, \end{aligned} \tag{12}$$

and

$$a^2 + b_1 b_2 = 1. \tag{13}$$

The form of Eqs. (79–11) permits the variables to be separated by cross-differentiation. The result is

$$\begin{aligned}\frac{d^2f}{d\xi^2} - f\left\{\frac{(\nu+1)(\nu+3)}{4\xi^2} - \frac{a(\nu+1)}{2\xi} + \frac{1}{4}\right\} &= 0,\\ \frac{d^2g}{d\xi^2} - g\left\{\frac{(\nu-1)(\nu+1)}{4\xi^2} - \frac{a(\nu+1)}{2\xi} + \frac{1}{4}\right\} &= 0.\end{aligned} \tag{14}$$

These equations both have the form of Whittaker's equation, Eq. (73–3); the identification is

$$k = \tfrac{1}{2}a(\nu+1) \tag{15}$$

and

$$l = \tfrac{1}{2}\nu + 1 \text{ for } f, \quad l = \tfrac{1}{2}\nu \text{ for } g. \tag{16}$$

Using Whittaker's functions, the general solution of Eqs. (79–13) is

$$\begin{aligned}f &= AW(k, \tfrac{1}{2}\nu+1, \xi) + BW(-k, \tfrac{1}{2}\nu+1, -\xi),\\ g &= CW(k, \tfrac{1}{2}\nu, \xi) + DW(-k, \tfrac{1}{2}\nu, -\xi),\end{aligned} \tag{17}$$

A, B, C and D being arbitrary constants. In order that these two functions shall also satisfy Eq. (79–11), the four constants must be restricted by the equations

$$\begin{aligned}(a+1)A &= b_1C,\\ b_2B &= (a+1)D,\end{aligned} \tag{18}$$

which can be derived from the recurrence formulae for Whittaker's functions.[(1),(3)]

There are thus just two arbitrary constants in the most general solution of Eqs. (79–11). These are further restricted by the requirement that the eigensolutions shall be regular at infinity (see Section 32). For large values of ξ, it is known (Ref. 1, p. 89) that

$$W(k, l, \xi) = \xi^k e^{-\frac{1}{2}\xi}\{1 + [l^2 - (k-\tfrac{1}{2})^2]/\xi + \cdots\}, \tag{19}$$

the series being asymptotically convergent. When $\omega^2 > \Omega^2$, all quantities are real, and therefore f and g will become infinite for large z unless $B = D = 0$. When $\omega^2 < \Omega^2$, on the other hand, ξ will be a pure imaginary, and f and g will remain finite only if $A = C = 0$. In either case, the eigensolutions contain just one arbitrary multiplicative constant.

The implications of Eq. (79–19) are more easily visualized when it is reformulated in terms of the amplitudes of the pressure and vertical displacement. Retracing the definitions, this reformulation is found to be

$$p_1 \sim z^{k - \frac{1}{2}(\nu + 1)} \mathrm{e}^{-\omega z/\alpha},$$
$$h_1 \sim z^{k + \frac{1}{2}\nu} \mathrm{e}^{-\omega z/\alpha}. \tag{20}$$

Both amplitudes therefore approach zero at high altitudes.

It remains to impose the boundary condition at the rigid surface of the atmosphere. This is

$$Q = 0 \quad \text{when } z = S \text{ or } \xi = \xi_S. \tag{21}$$

In case $\omega^2 > \Omega^2$, Eq. (79–10) transforms this into

$$b_1 W(k, \tfrac{1}{2}\nu + 1, \xi_S) \sin \tfrac{1}{2}\psi = (a + 1) W(k, \tfrac{1}{2}\nu, \xi_S) \cos \tfrac{1}{2}\psi. \tag{22}$$

This equation has no analogue in the case of the isothermal atmosphere. It is more analogous to the oceanic Eqs. (54–9) and (54–10), which arise because of the multiple reflection of the waves between the surface and bottom. Eq. (79–22) has a similar origin in the multiple reflection of the waves between the channel boundaries (Sections 77, 78).

As in the oceanic case, the numerical solution of Eq. (79–22) must await the further development of the theory of the Whittaker functions. However, it can be foreseen that the eigensolutions of this atmosphere will consist of sets, or modes, characterized by an ordinal number similar to that encountered in the oceanic case. As noted in the introduction, this is a major difference between the isothermal atmosphere and the present one.

80. The W-K-B approximation

As in Chapter XI, the computational difficulties of Eq. (79–22) cause one to resort to the W-K-B approximation for the introduction of the ordinal numbers. Again, the calculation will be based on the Eq. (74–4), which is

$$\int_{z'}^{z''} \gamma \, \mathrm{d}z = (m + \Delta)\pi. \tag{1}$$

In the case of those modes whose rays are not reflected, one will take $z' = z_-$, $z'' = z_+$ (cf. Eq. 77–4) and $m = |n|$, where n is the ordinal number. For simplicity, the approximation $\Delta = 0$ will be used, and the calculation will be made only for the case of no rotation. The equation connecting $\varkappa$, ω and n is then

$$\int_{z_-}^{z_+} \frac{\{-\varkappa^2 z^2 + 2r\varkappa z \cosh u - r^2 - s^2\}^{\frac{1}{2}}}{z} \, dz = |n|\pi. \tag{2}$$

This is an elementary integral, whose evaluation leads to

$$\pi\{r \cosh u - (r^2 + s^2)^{\frac{1}{2}}\} = |n|\pi. \tag{3}$$

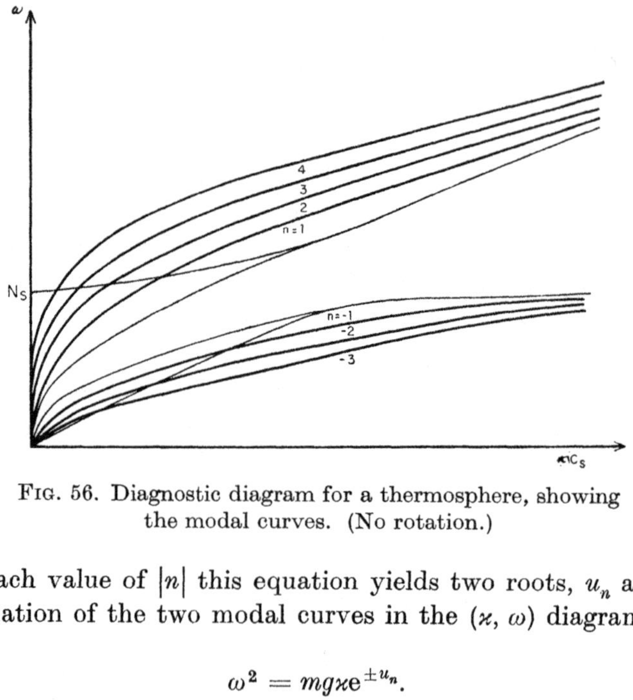

FIG. 56. Diagnostic diagram for a thermosphere, showing the modal curves. (No rotation.)

For each value of $|n|$ this equation yields two roots, u_n and $-u_n$. The equation of the two modal curves in the $(\varkappa, \omega)$ diagram is thus

$$\omega^2 = mg\varkappa e^{\pm u_n}. \tag{4}$$

The modal curves are parabolae, so long as they remain in the doubly-shaded area of Fig. 55. This is shown, for several values of n, on Fig. 56.

In the case of those modes whose rays are reflected at the ground, Eq. (80–2) becomes

$$\int_{S}^{z_+} \frac{\{\ldots\}^{\frac{1}{2}}}{z} \, dz = |n|\pi. \tag{5}$$

The integral is still elementary, but leads to the more complicated formula

$$-a \cos \theta_S + b\left(\frac{\pi}{2} - \theta_S\right) - 2(r^2 + s^2)^{\frac{1}{2}} \tan^{-1}\left\{\left[\frac{b + a}{b - a}\right]^{\frac{1}{2}}\right\}$$

$$- \tan^{-1}\left\{\frac{a + b \tan \frac{1}{2}\theta_S}{(r^2 + s^2)^{\frac{1}{2}}}\right\} = |n|\pi, \quad (6)$$

where

$$b = r \cosh u,$$
$$\varkappa S = b - a \sin \theta_S, \quad (7)$$

and the other quantities have the same significance as before. The graphs of Eq. (80–6), for various values of n, are not parabolae. Their general course has been sketched on Fig. 56.

For sufficiently small values of $\varkappa$, each of the modal curves corresponds to rays that are not reflected at the ground. Over this portion of the curve, it is a parabola, as shown by Eq. (80–4). For larger values of $\varkappa$, each modal curve corresponds to rays that are reflected at the ground, as shown by Eq. (80–6). The most striking difference between this $(\varkappa, \omega)$ diagram and that of the oceans, is the fact that the acoustic, as well as the gravity modes, originate at $\varkappa = 0$, $\omega = 0$. This is caused by the infinite value of c at $z = +\infty$.

References for Chapter XIII

1. W. Magnus and F. Oberhettinger, *Formulas and Theorems for the Special Functions of Mathematical Physics*, Chelsea, New York, 1949.
2. L. Schlesinger, *Einführung in die Theorie der gewöhnlichen Differentialgleichungen*, de Gruyter, Berlin, 1922.
3. E. T. Whittaker and G. N. Watson, *A Course of Modern Analysis*, Cambridge Univ. Press, 1927.

CHAPTER XIV

GENERAL THEORY OF THE RESIDUAL EQUATIONS

81. Introduction

In Chapter VII, it was shown that the eigensolutions depend on the altitude in a way that is essentially the same in all cases. The equations involved have the form

$$c\left(\frac{\mathrm{d}P}{\mathrm{d}z} + \Gamma P\right) = (N^2 - \nu)Q,$$
$$c\left(\frac{\mathrm{d}Q}{\mathrm{d}z} - \Gamma Q\right) = (1 - c^2/\mu)P. \tag{1}$$

If the level surfaces are plane or spherical and there is no rotation,

$$\nu = \omega^2, \qquad \mu = \omega^2/\varkappa^2. \tag{2}$$

If the level surfaces are plane, and the traditional approximation, $\Omega_H = 0$, is made, then

$$\nu = \omega^2, \qquad \mu = (\omega^2 - \Omega_V{}^2)/\varkappa^2. \tag{3}$$

If the level surfaces are spheres, and the traditional approximation is used, one again has Eq. (81–2) provided $\varkappa^2$ is defined as the eigenvalue of Laplace's tidal equation, Eq. (43–12).

Special cases of Eq. (81–1) have been treated in the previous chapters. It has been seen that even quite simple physical assumptions concerning the stratification lead to mathematical problems that are not easy. In Section 72, however, a quite powerful method for treating the case $\Gamma = 0$, $c =$ const., was developed. It is true that this method was not quantitative, but it did yield important qualitative results. This chapter will be devoted to a generalization of that method.

In Section 72, the first step was the elimination of P from the first equation, leading to Eq. (72–4), which is a second-order equation for Q. If this same procedure is attempted in more general cases, the result is an analytically-complicated equation for Q. Moreover, from a mathematical point of view, it must be classed as "an exceptional case of an exceptional case." One must therefore seek to develop methods for studying Eq. (81–1) directly.

The method used in Section 72 is based on a few general theorems, due to Sturm and Liouville. No formal proof of these theorems was given, but instead, the reader's attention was directed to their geometric content and this was used to prove the needed results. Formal proofs of these theorems are to be found elsewhere.[1]

The plan will be, first to transform Eq. (81–1) to a standard or canonic form, then to prove a few general theorems concerning these differential equations, and finally, to apply these theorems to special cases. In particular, it will be shown that they provide considerable new information concerning the thermosphere problem of Chapter XIII.

82. Canonic form of the residual equations

The canonic variables. Instead of the variables P, Q it will be convenient to use

$$\begin{aligned} \mathscr{P} &= (P/\rho_S g) \exp\left\{+\int \Gamma \, dz\right\}, \\ \mathscr{Q} &= Q \exp\left\{-\int \Gamma \, dz\right\}, \end{aligned} \tag{1}$$

where ρ_S is a constant, with the dimensions of density, which is conveniently identified with the surface density of the air or water. $\mathscr{Q}$ and $\mathscr{P}$ are then measured in centimetres.

The variables $\mathscr{P}$ and $\mathscr{Q}$ may also be related to the c.g.s. variables p_1, and η_1, used in Chapters II and III. In Section 22, P and Q were defined as

$$P = p_1(\rho_0 c_0)^{-\frac{1}{2}}, \qquad Q = (\eta_1/\eta_0')(\rho_0 c_0)^{+\frac{1}{2}}, \tag{2}$$

and in Section 23,

$$\Gamma = \frac{1}{2}\frac{(\rho_0 c_0)'}{(\rho_0 c_0)} + \frac{g}{c_0^{\,2}}. \tag{3}$$

It readily follows that

$$\begin{aligned}\mathscr{P} &= (p_2/\rho_S g)\exp(+\delta),\\ \mathscr{Q} &= (\eta_1/\eta_0')\exp(-\delta),\end{aligned} \tag{4}$$

where

$$\delta = g\int dz/c_0^{\,2}. \tag{5}$$

In Section 9, it was also shown that the vertical displacement of the fluid is

$$h_1 = -\eta_1/\eta_0', \tag{6}$$

so that

$$\mathscr{Q} = -h_1\exp(-\delta). \tag{7}$$

The canonic equations. On substituting for P and Q in terms of $\mathscr{P}$ and $\mathscr{Q}$, the residual equations, Eq. (81–1) take the form

$$\begin{aligned}\frac{d\mathscr{Q}}{dz} &= A(z,\mu)\mathscr{P},\\ \frac{d\mathscr{P}}{dz} &= -B(z,\nu)\mathscr{Q},\end{aligned} \tag{8}$$

where

$$A = a\left(\frac{1}{c^2} - \frac{1}{\mu}\right), \tag{9}$$

$$B = \frac{1}{a}(\nu - N^2), \tag{10}$$

$$a = (\rho_S g/\rho_0)\exp(-2\delta). \tag{11}$$

The factor a is positive for all values of z. Just as in Section 72, one was concerned largely with the sign of the convexity, so one will now be concerned largely with the signs of A and B, and these are independent of a.

The function

$$H = \tfrac{1}{2}(A\mathscr{P}^2 + B\mathscr{Q}^2) \tag{12}$$

is known as the Hamiltonian of Eq. (82–8), which may also be written

$$\frac{d\mathscr{Q}}{dz} = \frac{\partial H}{\partial\mathscr{P}},\qquad \frac{d\mathscr{P}}{dz} = -\frac{\partial H}{\partial\mathscr{Q}}. \tag{13}$$

Constant coefficients. If the coefficients A and B are independent of z, the Eqs. (82–8) or (82–13) become very simple. An examination of this case furnishes a preview of the general theorems that follow. On differentiating Eq. (82–12),

$$\frac{\mathrm{d}H}{\mathrm{d}z} = \frac{\partial H}{\partial \mathscr{Q}}\frac{\mathrm{d}\mathscr{Q}}{\mathrm{d}z} + \frac{\partial H}{\partial \mathscr{P}}\frac{\mathrm{d}\mathscr{P}}{\mathrm{d}z},$$

because $\mathrm{d}A/\mathrm{d}z = \mathrm{d}B/\mathrm{d}z = 0$. Comparing this with Eq. (82–11), it is seen that

$$\mathrm{d}H/\mathrm{d}z = 0$$

or

$$A\mathscr{P}^2 + B\mathscr{Q}^2 = \text{const.} \tag{12}$$

This equation may be plotted in a ($\mathscr{P}$, $\mathscr{Q}$) diagram; if A and B have the same sign, its graph is an ellipse, and if they have opposite signs, its graph is an hyperbola. The dominant role of the signs of A and B is thus established even in this simplest case.

The phase diagram. The ($\mathscr{P}$, $\mathscr{Q}$) diagram will be very useful in the following sections. It is commonly known as the phase-space or phase diagram. Recalling that $\mathscr{Q}$ is entropy and $\mathscr{P}$ is pressure (in appropriate units) the phase diagram is seen to be essentially the entropy–pressure diagram of thermodynamic theory (Section 7). A special case of the phase diagram has already been encountered in Section 47, Fig. 19.

In the theory of one-dimensional mechanics, equations are studied that have the same canonic form as Eq. (82–11). There, however, the independent variable is not the altitude, but the time. One therefore imagines $\mathscr{P}$ and $\mathscr{Q}$ as the co-ordinates of a point that moves in phase space and traces a curve. It is convenient to use this terminology in the present case; this can be justified by imagining z to increase linearly with some fictitious time. One then arrives at the idea of a moving phase point whose rectangular co-ordinates are $\mathscr{P}$ and $\mathscr{Q}$, and which traces a path in the phase diagram.

83. General theorems concerning the phase paths

In this section, the parameters μ and ν will be supposed to have constant values and the dependence of $\mathscr{P}$ and $\mathscr{Q}$ on z will be examined. The quantities A and B are to be finite and continuous functions of z

for all values of this variable that come under consideration. Theorem I will not be proven here; the methods for constructing its proof are given elsewhere.[2]

Theorem I

The phase paths are continuous curves. Let $\mathscr{P}_0, \mathscr{Q}_0$ *be any point of the phase diagram other than the origin, and* z_0 *an arbitrary finite value of the altitude; then there is exactly one phase path that passes through* $\mathscr{P}_0, \mathscr{Q}_0$ *at* z_0. *No phase path goes through the origin for any finite value of* z, *but some paths may approach the origin if* $z \rightarrow \pm \infty$.

The canonic equations show that if z is a root of $\mathscr{P}$, it is also a root of $\mathrm{d}\mathscr{Q}/\mathrm{d}z$; and if it is a root of $\mathscr{Q}$, it is also a root of $\mathrm{d}\mathscr{P}/\mathrm{d}z$. This may be formulated as

Theorem II

All phase paths intersect the co-ordinate axes at right angles.

Similarly, if z is a root of A, it is also a root of $\mathrm{d}\mathscr{Q}/\mathrm{d}z$; and if it is a root of B, $\mathrm{d}\mathscr{P}/\mathrm{d}z = 0$. Hence:

Theorem III

If z *is a root of* A (*or* B), *then the phase path has a horizontal* (*or vertical*) *tangent for this value of* z.

Moreover, if z is finite, the tangent of a phase path can be horizontal or vertical only under the conditions of these theorems.

A most important conclusion is

Liouville's Theorem

If $\mathscr{P}_1, \mathscr{Q}_1$ *and* $\mathscr{P}_2, \mathscr{Q}_2$ *are two solutions of the canonic equations, the determinant*

$$L_{12} = \begin{vmatrix} \mathscr{P}_1 & \mathscr{Q}_1 \\ \mathscr{P}_2 & \mathscr{Q}_2 \end{vmatrix}$$

is independent of z.

For,

$$\frac{\mathrm{d}L_{12}}{\mathrm{d}z} = \frac{\mathrm{d}\mathscr{P}_1}{\mathrm{d}z}\mathscr{Q}_2 + \mathscr{P}_1\frac{\mathrm{d}\mathscr{Q}_2}{\mathrm{d}z} - \frac{\mathrm{d}\mathscr{P}_2}{\mathrm{d}z}\mathscr{Q}_1 - \mathscr{P}_2\frac{\mathrm{d}\mathscr{Q}_1}{\mathrm{d}z}.$$

Substituting from Eq. (82–8), this becomes

$$\frac{\mathrm{d}L_{12}}{\mathrm{d}z} = -B\mathcal{Q}_1\mathcal{Q}_2 + A\mathcal{P}_1\mathcal{P}_2 + B\mathcal{Q}_2\mathcal{Q}_1 - A\mathcal{P}_2\mathcal{P}_1 = 0. \qquad (1)$$

An important corollary is

Theorem IV

If $\mathcal{P}_1$, $\mathcal{Q}_1$, $\mathcal{P}_2$, $\mathcal{Q}_2$ and $\mathcal{P}_3$, $\mathcal{Q}_3$ are solutions of the canonic equations, and $L_{12} \neq 0$; then there are two constants, α and β, such that

$$\begin{aligned} \mathcal{P}_3 &= \alpha\mathcal{P}_1 + \beta\mathcal{P}_2, \\ \mathcal{Q}_3 &= \alpha\mathcal{Q}_1 + \beta\mathcal{Q}_2. \end{aligned} \qquad (2)$$

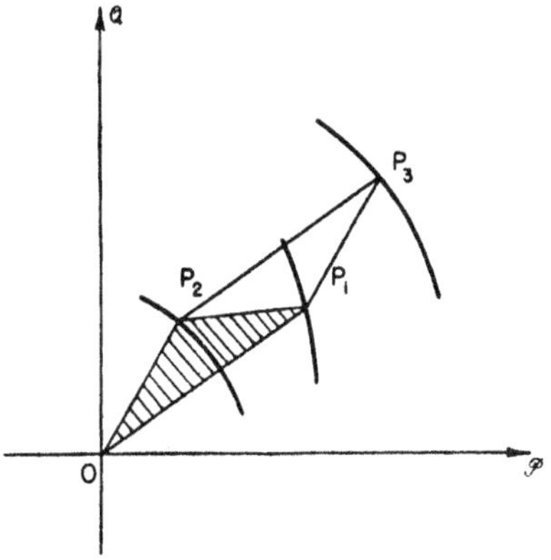

FIG. 57. Illustrating Liouville's Theorem.

This is proven by solving the equations for α and β, and noting that

$$\alpha = -L_{23}/L_{12}, \quad \beta = -L_{31}/L_{12},$$

which are constants by Liouville's theorem.

Liouville's theorem has a simple geometric interpretation; let P_i be the point $\mathcal{P}_i(z)$, $\mathcal{Q}_i(z)$, and consider simultaneous positions of P_1 and P_2. Then L_{12} is twice the area of the shaded triangle in Fig. 57, and as the points describe their paths, this area remains constant. If $L_{12} > 0$, the circuit OP_1P_2O is described in the positive or counter-clockwise sense; if $L_{12} < 0$, the circuit is described in the negative sense. Theorem IV states, among other things, that the path of P_3, on Fig. 57, which is constructed by addition of the moving vectors OP_1

and OP_2, is again a phase path. This is analogous to the vector diagrams used by electrical engineers.

Thus far, the theorems have not referred to the signs of A and B; these enter the discussion when polar co-ordinates are introduced in the phase space: let

$$\mathscr{P} = \mathscr{R} \cos \vartheta, \quad \mathscr{Q} = \mathscr{R} \sin \vartheta, \quad \mathscr{R} \geqslant 0, \tag{3}$$

then the canonic equations become

$$\frac{d\vartheta}{dz} = A \cos^2 \vartheta + B \sin^2 \vartheta, \tag{4}$$

$$\frac{1}{\mathscr{R}} \frac{d\mathscr{R}}{dz} = (A - B) \sin \vartheta \cos \vartheta. \tag{5}$$

The importance of the roots of A and B has already appeared in Theorems II and III; they divide the z-axis (and therefore also the phase paths) into segments of four types, which may be denoted by $(++)$, $(--)$, $(+-)$ and $(-+)$, the significance of these symbols being

$$(++)\colon\ A > 0,\ B > 0;$$

$$(--)\colon\ A < 0,\ B < 0,$$

$$(+-)\colon\ A > 0,\ B < 0;$$

$$(-+)\colon\ A < 0,\ B > 0.$$

Such symbols will be called the signatures of the segments. Two theorems follow at once from these definitions and Eqs. (83–4) and (83–5):

Theorem V

In $(++)$ segments, the angular velocity of the phase point is positive; in $(--)$ segments it is negative.

Theorem VI

In $(+-)$ segments, the phase point moves away from the origin if in the first or third quadrants of the diagram, and toward the origin if in the second or fourth. In $(-+)$ segments, the directions of motion will be reversed.

These theorems are illustrated by Figs. 58 and 59, to which frequent reference will be made below. They have numerous corollaries, such as: In $(+-)$ or $(-+)$ segments, at most one of the functions $\mathscr{P}$ and $\mathscr{Q}$ can have a root; if one function has a root, the other has an extremum. In$(++)$ or $(--)$ segments, both $\mathscr{P}$ and $\mathscr{Q}$ may have many roots, and their extrema are separated by roots.

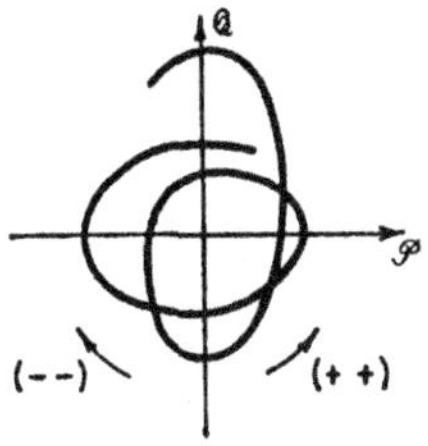

FIG. 58. Sketch of phase path in a $(++)$ or $(--)$ segment; the direction of the moving phase point in the two cases is indicated by the arrows.

These results are analogous to those of Section 72. The functions $\mathscr{P}$ and $\mathscr{Q}$ oscillate in $(++)$ or $(--)$ segments; they do not oscillate in $(+-)$ or $(-+)$ segments. But, whereas the oscillatory or non-oscillatory character of the solution was determined by the sign of the convexity in Section 72, here the signs of both A and B are needed for this determination.

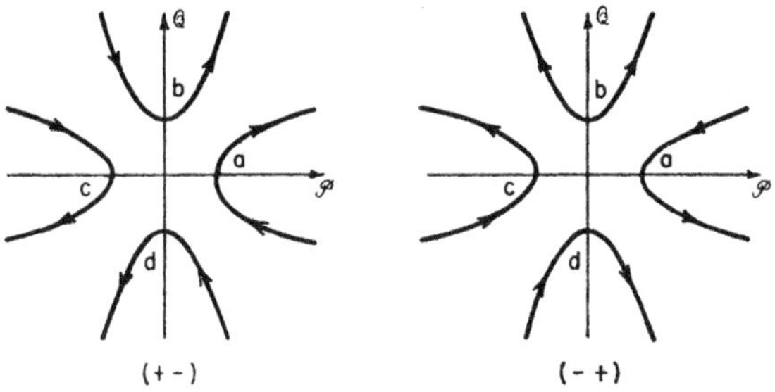

FIG. 59. Sketches of phase paths in $(+-)$ and $(-+)$ segments.

If one of the segments extends to infinity, several new phenomena may appear. If the segment is of the oscillatory, $(++)$ or $(--)$, type, the angular velocity will vanish only if

$$\lim_{z\to\infty} A,\, B = 0.$$

In other cases, the phase path will cross every axis an infinite number of times, and the logarithm of its distance from the origin may become infinite with z. If this infinity is negative, the path will spiral toward the origin without, however, passing through it. If the infinity is positive, the phase point will recede to infinity along a spiral path. If A and B approach finite limits the path will become asymptotic to an ellipse. Other possibilities exist, and each case must be investigated in detail.

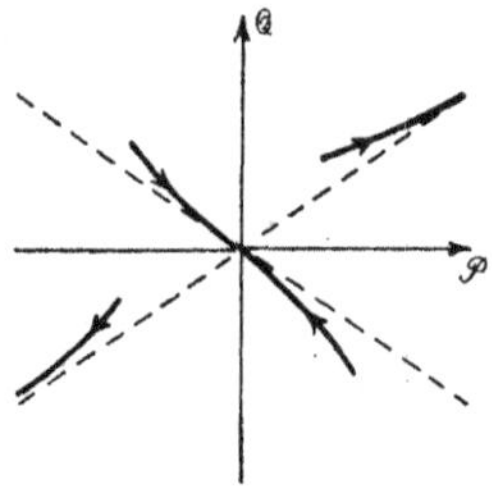

FIG. 60. Sketches of the asymptotic behaviour of phase paths in (+ −) or (− +) segments.

If the infinite segment is of the non-oscillatory, (+ −) or (− +), type and extends to $z = +\infty$, the angle defined by

$$\tan^2 \vartheta_a = \lim_{z \to +\infty} (-A/B) \tag{6}$$

will be important. The phase paths will have the asymptotic characters shown by Fig. 60, the dotted lines making the angles $\pm\vartheta_a$ with the $\mathscr{P}$-axis. In (+ −) segments, the phase point will recede from the origin in the direction ϑ_a, provided it is in the first or third quadrants. If it is in the second or fourth quadrant, it will approach, but not pass through, the origin in the direction $-\vartheta_a$. For (− +) segments, the situation is reversed. If the segment extends, not to $z = +\infty$ but to $z = -\infty$, the directions of the arrows are reversed.

The manner in which the segments of a phase path are joined together is governed by Theorem III. The symbolism (+ +) (− +) (− −), etc., will be convenient for designating a sequence of segments separated by roots of A and B. This example indicates, in an obvious way, that A and B are initially positive, but change sign in succession as z increases. In the same way, (+ +) (− +) (+ +) indicates that A has two roots, B none. A possible phase path for each of these two cases is shown by Fig. 61. The reader will have no difficulty sketching others.

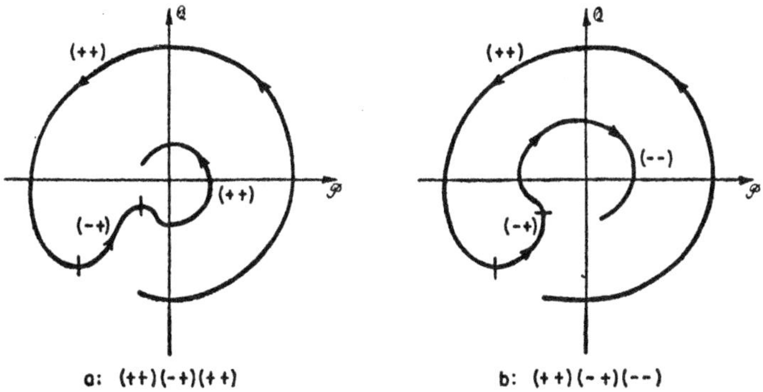

FIG. 61. Sketches of phase paths with compound signatures.

84. Sturm's comparison theorem

Sturm's formula. In the preceding section, some geometric properties of the phase paths were established. Additional results are obtained by comparing the phase paths determined by two different systems of canonic equations:

$$\frac{d\mathscr{P}_1}{dz} = -B_1\mathscr{Q}_1, \qquad \frac{d\mathscr{Q}_1}{dz} = A_1\mathscr{P}_1;$$
$$\frac{d\mathscr{P}_2}{dz} = -B_2\mathscr{Q}_2, \qquad \frac{d\mathscr{Q}_2}{dz} = A_2\mathscr{P}_2. \tag{1}$$

The Liouville determinant

$$L_{12} = \mathscr{P}_1\mathscr{Q}_2 - \mathscr{P}_2\mathscr{Q}_1 \tag{2}$$

can be formed, and the generalization of Liouville's theorem is

$$\frac{dL_{12}}{dz} = (A_2 - A_1)\mathscr{P}_1\mathscr{P}_2 + (B_2 - B_1)\mathscr{Q}_1\mathscr{Q}_2. \tag{3}$$

This equation is known as Sturm's formula, and is the source of many theorems.(1)

In applications, matters are usually arranged so that $L_{12} = 0$ at a certain value of z—say z_0—and that $A_2 > A_1$, $B_2 > B_1$ for an interval that includes z_0. If P_1 and P_2 are then in the same quadrant during this interval, the right-hand side of Eq. (84–3) is positive throughout that interval, and $L_{12} > 0$ for $z > z_0$, while $L_{12} < 0$ for $z < z_0$. Since

$$L_{12} = \mathscr{R}_1 \mathscr{R}_2 \sin(\vartheta_2 - \vartheta_1),$$

various geometric conclusions follow.

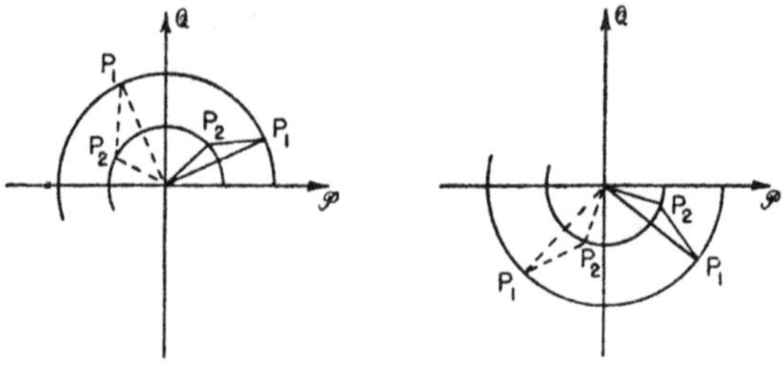

FIG. 62. Illustrating the oscillation theorem.

The oscillation theorems. For example, suppose that $A_2 > A_1 > 0$, $B_2 > B_1 > 0$ for all $z > 0$, and that

$$\mathscr{P}_1 = \mathscr{P}_2 = 1, \qquad \mathscr{Q}_1 = \mathscr{Q}_2 = 0$$

at $z = 0$. Then the segment $z > 0$ is $(++)$, and both points P_1 and P_2 will have positive angular velocities $d\vartheta_1/dz$, $d\vartheta_2/dz$. For some interval $0 < z < z'$, both points will be in the first quadrant, and the right-hand side of Eq. (84–3) will be positive; during this interval $L_{12} > 0$, or $\vartheta_2 > \vartheta_1$ (Fig. 62). This conclusion remains valid until P_2 crosses the $\mathscr{Q}$-axis: then $\mathscr{P}_2\mathscr{P}_1$ becomes negative and the conclusion might be invalidated. But it has at least been established that P_2 enters the second quadrant before P_1.

When P_1 enters the second quadrant, P_2 may still be in that quadrant, or in some later quadrant. In the second case, it will surely have entered the third quadrant before P_1; but it can be shown that this is true even in the first case. For then $L_{12} > 0$ as P_1 enters the second quadrant, and cannot decrease until P_2 leaves that quadrant. Since $L_{12} > 0$ as P_2 leaves the second quadrant, this must occur before P_1 leaves that quadrant.

Continuing in this manner, it can be established by induction that if P_1 has crossed the $\mathscr{P}$- (or $\mathscr{Q}$-) axis n times in the interval $0 \to z$, P_2 will have crossed that axis at least n times and perhaps oftener. This conclusion is usually summarized by saying that P_2 oscillates more rapidly than P_1, and is called an oscillation theorem.

There are a large number of oscillation theorems, differing in their hypotheses and conclusions, but proven in a manner similar to the above. Fig. 62 also illustrates the corresponding theorem for a (— —) segment in which $0 > A_2 > A_1$, $0 > B_2 > B_1$ for $z > 0$. In this case, P_1 oscillates more rapidly than P_2.

Dependence on the parameters. If A and B are given by Eq. (82–9), and A_2, B_2 differ from A_1, B_1 only because the parameters μ and ν have different values, Sturm's formula becomes

$$\frac{\mathrm{d}L_{12}}{\mathrm{d}z} = a\left(\frac{1}{\mu_1} - \frac{1}{\mu_2}\right)\mathscr{P}_1\mathscr{P}_2 + \frac{1}{a}(\nu_2 - \nu_1)\mathscr{Q}_1\mathscr{Q}_2. \tag{4}$$

This formula will be applied to the case of a rigid boundary, so that

$$\mathscr{Q}_1 = \mathscr{Q}_2 = 0 \quad \text{at} \quad z = 0\,. \tag{5}$$

For convenience, the solutions will be normalized by the additional equations

$$\mathscr{P}_1 = \mathscr{P}_2 = 1 \quad \text{at} \quad z = 0\,. \tag{6}$$

Then $L_{12} = 0$ at $z = 0$ and Eq. (84–4) integrates to

$$L_{12} = \left(\frac{1}{\mu_1} - \frac{1}{\mu_2}\right)\int_0^z a\,\mathscr{P}_1\mathscr{P}_2\,\mathrm{d}z + (\nu_2 - \nu_1)\int_0^z (\mathscr{Q}_1\mathscr{Q}_2/a)\,\mathrm{d}z. \tag{7}$$

It will be supposed throughout the following, that $z > 0$; then both integrals in Eq. (84–7) will be positive provided neither $|\mu_2 - \mu_1|$ nor $|\nu_2 - \nu_1|$ is too great. For these integrals are certainly positive when $\mu_2 = \mu_1$ and $\nu_2 = \nu_1$—which implies $\mathscr{P}_1 = \mathscr{P}_2$ and $\mathscr{Q}_1 = \mathscr{Q}_2$ by Theorem I. It can also be proven that $\mathscr{P}$, $\mathscr{Q}$ are continuous functions of μ and ν. This continuity then guarantees against an abrupt change of sign when μ_1, ν_1 is not exactly the same as μ_2, ν_2.

Consequently L_{12} will be positive when $\mu_2 - \mu_1 = \delta\mu > 0$, and $\nu_2 - \nu_1 = \delta\nu > 0$, provided that $\delta\mu$ and $\delta\nu$ are not too large. Since

$$\begin{aligned} L_{12} &= \mathscr{R}_1\mathscr{R}_2 \sin(\vartheta_2 - \vartheta_1) \\ &= \mathscr{R}_1\mathscr{R}_2 \sin(\delta\vartheta), \end{aligned}$$

it follows that $\delta\vartheta > 0$ under these conditions. In other words, $\vartheta(z, \mu, \nu)$ is a monotonically-increasing function of both μ and ν for all positive values of z. It is important to note that this is true for segments of any type. It is therefore a very powerful form of oscillation theorem, but applies only under the conditions explicitly stated above.

85. The W-K-B approximation

These results can be derived more concisely, but less rigorously, by using the W-K-B approximation. Let

$$\mathscr{P} = \mathscr{P}_0 \cos \phi, \quad \mathscr{Q} = \mathscr{Q}_0 \sin \phi. \tag{1}$$

It is to be noted that the angles ϕ and ϑ are not identical, since it is not to be supposed that $\mathscr{P}_0$ and $\mathscr{Q}_0$ are equal. Substitution of this equation into

$$\frac{d\mathscr{Q}}{dz} = A\mathscr{P}, \qquad \frac{d\mathscr{P}}{dz} = -B\mathscr{Q}, \tag{2}$$

leads to

$$\begin{aligned} \frac{d\mathscr{Q}_0}{dz} \sin \phi + \mathscr{Q}_0 \cos \phi \frac{d\phi}{dz} &= A\mathscr{P}_0 \cos \phi, \\ \frac{d\mathscr{P}_0}{dz} \cos \phi - \mathscr{P}_0 \sin \phi \frac{d\phi}{dz} &= -B\mathscr{Q}_0 \sin \phi. \end{aligned} \tag{3}$$

The essence of the W-K-B approximation consists in neglecting the derivatives of $\mathscr{P}_0$ and $\mathscr{Q}_0$ in comparison to that of ϕ. To this approximation, therefore

$$\begin{aligned} \mathscr{Q}_0 \frac{d\phi}{dz} &= A\mathscr{P}_0, \\ \mathscr{P}_0 \frac{d\phi}{dz} &= B\mathscr{Q}_0. \end{aligned} \tag{4}$$

Eliminating $\mathscr{P}_0$ and $\mathscr{Q}_0$, it is seen that

$$\left(\frac{d\phi}{dz}\right)^2 = AB. \tag{5}$$

Using Eq. (82–9),

$$\left(\frac{d\phi}{dz}\right)^2 = \left(\frac{1}{c^2} - \frac{1}{\mu}\right)(\nu - N^2). \tag{6}$$

It is at once seen that ϕ can be real only in $(++)$ or $(--)$ segments, in conformity with the oscillatory character of the solutions established rigorously in the previous sections. It is also seen that $d\phi/dz$ is a monotonically-increasing function of both μ and ν.

In order to obtain a similar approximation for $(+-)$ and $(-+)$ segments, Eq. (85–1) may be altered to

$$\mathscr{P} = \mathscr{P}_0 \cosh \phi, \qquad \mathscr{Q} = \mathscr{Q}_0 \sinh \phi. \tag{7}$$

Then

$$\begin{aligned} \frac{d\mathscr{Q}_0}{dz} \sinh \phi + \mathscr{Q}_0 \cosh \phi \frac{d\phi}{dz} &= A \mathscr{P}_0 \cosh \phi, \\ \frac{d\mathscr{P}_0}{dz} \cosh \phi + \mathscr{P}_0 \sinh \phi \frac{d\phi}{dz} &= -B \mathscr{Q}_0 \sinh \phi, \end{aligned} \tag{8}$$

and, if one again neglects the derivatives of $\mathscr{P}_0$ and $\mathscr{Q}_0$, the equation

$$\left(\frac{d\phi}{dz}\right)^2 = -AB \tag{9}$$

results. Hence, with this definition, ϕ can be real only in $(+-)$ or $(-+)$ segments, and the non-oscillatory nature of the solution in such segments is established.

References for Chapter XIV

1. M. Bôcher, *Leçons sur la Methodes de Sturm*, Gauthier-Villars, Paris, 1917.
2. R. Bellman, *Stability Theory of Differential Equations*, McGraw-Hill, New York, 1953.

CHAPTER XV

APPLICATIONS OF THE GENERAL THEORY

86. The thermosphere

CHAPTER XIII was devoted to the theory of an atmosphere with a constant, positive temperature gradient. It was possible to deduce the general features of the diagnostic diagram, which shows those values of $\varkappa$ and ω that are associated with eigensolutions. In this section, the general theorems proven in Chapter XIV will be used to deduce some of the properties of the eigensolutions. The results will be analogous to those deduced in Section 72 for the thermocline.

It was seen that

$$N^2 = g\gamma r^2/z\nu, \qquad c^2 = g\gamma z/\nu \tag{1}$$

for an atmosphere of this kind. These two functions are plotted on Fig. 63. Since

$$\begin{aligned} A &= a\left(\frac{1}{c^2} - \frac{\varkappa^2}{\omega^2}\right) \\ &= \frac{a\varkappa^2}{\omega^2 c^2}\left(\frac{\omega^2}{\varkappa^2} - c^2\right), \end{aligned} \tag{2}$$

$$B = \frac{1}{a}(\omega^2 - N^2), \tag{3}$$

the horizontal lines $c^2 = \omega^2/\varkappa^2$ and $N^2 = \omega^2$ are also shown on these graphs. As ω and $\varkappa$ vary, these lines, of course, change their positions. The root of A is at $z = z_A$, that of B, at $z = z_B$. When $z < z_A$, the function A is positive; when $z < z_B$, the function B is negative. Hence the positive z-axis is divided into three segments by the two roots and in the case illustrated, has the signature $(+-)$ $(++)$

$(-+)$. Moreover, in the notation of Chapter XIII, $z_A \sim z_+$, $z_B \sim z_-$. For the values of ω and $\varkappa$ illustrated by these graphs, one is therefore dealing with an acoustic ray that does not intersect the surface of the planet, but runs its entire course aloft, essentially in the $(++)$ segment. Moreover, it follows from the general theorems of Chapter

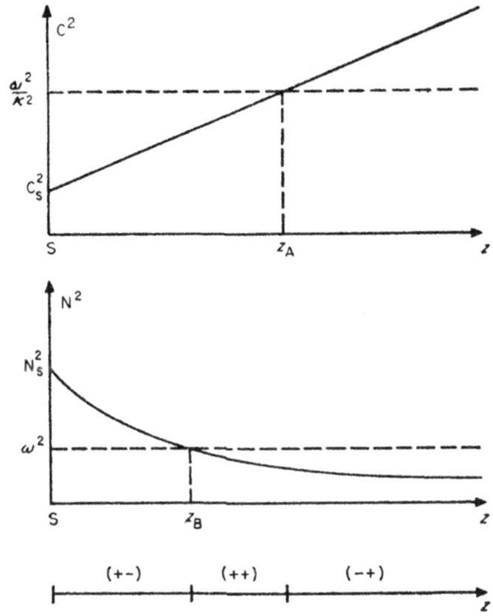

FIG. 63. Determination of the signature for given values of ω^2 and $\omega^2/\varkappa^2$.

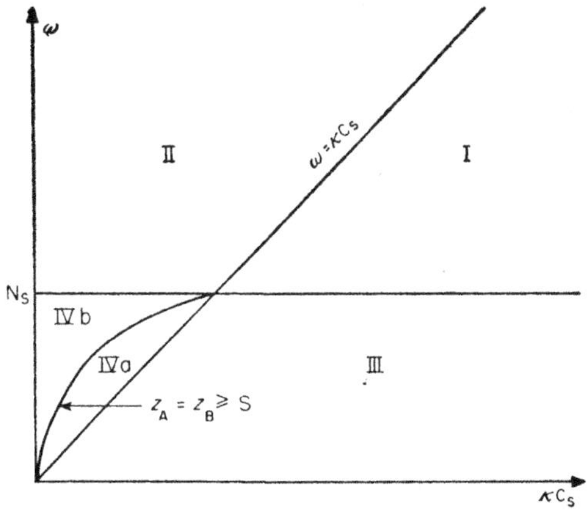

FIG. 64. Subdivision of the diagnostic diagram for a thermosphere; in any one numbered area, all phase paths have a single signature, as shown in the table.

XIV that the ($\mathscr{P}$, $\mathscr{Q}$) fields oscillate in the $(++)$ segment, and do not oscillate in the other two.

These qualitative conclusions concerning the eigensolutions will be made more precise below; first, however, the various possible signatures will be investigated, and transferred to the diagnostic diagram of Fig. 64. On this diagram, the lines $\omega = N_S$, $\omega = \varkappa c_S$ and the curve $z_A = z_B$ have been entered. They divide the $(\varkappa, \omega)$ plane into five areas, each of which is characterized by a different signature, as indicated below. It will be noted that all five end with the $(-+)$ signature that is characteristic of the thermosphere.

		$(-+)$	$z_A < S,\ z_B < S,$	area I
	$(++)$	$(-+)$	$z_A > S,\ z_B < S,$	area II
$(+-)$	$(++)$	$(-+)$	$z_A > z_B > S,$	area IV[b]
	$(--)$	$(-+)$	$z_A < S,\ z_B > S,$	area III
$(+-)$	$(--)$	$(-+)$	$z_B > z_A > S,$	area IV[a]

The grouping of the signatures in this table follows an obvious system, and provides a convenient order in which to consider the phase paths. These must all start on the $\mathscr{P}$-axis, since the boundary condition at the rigid lower surface of the atmosphere implies $\mathscr{Q} = 0$ at $z = S$.

Area I

Since the signature in this area is $(-+)$, the general character of the phase path must be as described in Theorem V and shown by Figs. 59 and 60. A further item of information is that the asymptotic angle, ϑ_a, is $\pi/2$; for

$$\tan^2 \vartheta_a = \lim_{z \to \infty} (-A/B)$$

and $-A$ increases indefinitely with z, while B approaches ω^2/a. Hence the phase path is non-oscillatory, and runs its entire length in the fourth quadrant of the phase diagram. As $z \to \infty$, both $\mathscr{P}$ and $\mathscr{Q}$ become infinite, and $\mathscr{P}$ ultimately becomes much greater than $\mathscr{Q}$.

However, the eigensolutions must remain finite for all values of z.

Hence one concludes that no eigensolutions correspond to points of area I. This extends the conclusion of Section 77, that there are no rays which correspond to points in this area.

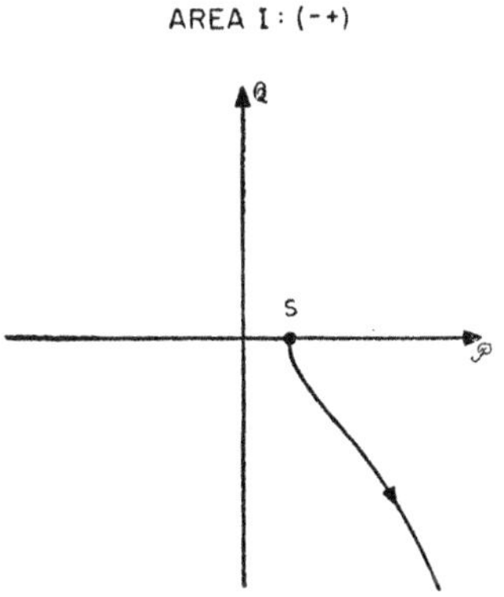

FIG. 65. Sketch of a phase path for AREA I, showing that it cannot be an eigensolution.

AREA II

Since the signature is (++) (−+), the phase paths begin as spirals described with positive angular velocity (Theorem IV). At

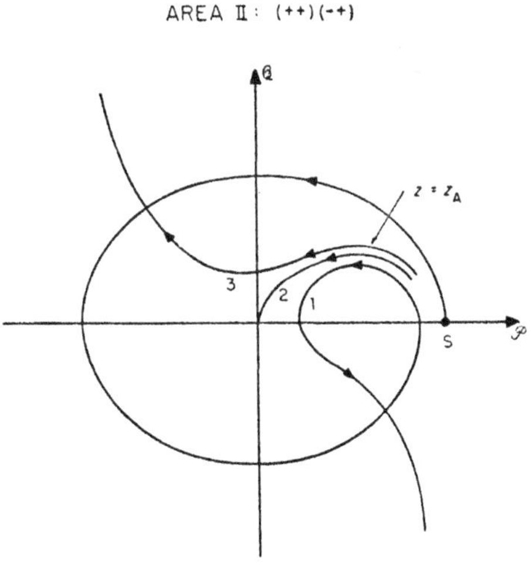

FIG. 66. Sketch of phase paths for AREA II, showing the distinction between eigensolutions (curve 2) and other solutions (curves 1 and 3).

$z = z_A$, the coefficient A vanishes, and the path must have a horizontal tangent (Theorem III). For larger values of z, they have the characteristics described in Theorem V, and Fig. 59. The asymptotic angle ϑ_a is again $\pi/2$.

Hence the general character of the phase paths is as shown on Fig. 66. First consider the curve marked 1 as z increases from zero. The phase point describes a roughly elliptic spiral, making somewhat more than one turn. When $z = z_A$, the phase point has returned to the first quadrant, and the curve has a horizontal tangent (Theorem III). Thus far, the curve is described by Theorem IV, since the segment is $(++)$. As z increases beyond z_A, it enters the $(-+)$ segment, and the curve is described by Theorem V. Curve 1 shows the case in which the phase path crosses the $\mathscr{P}$-axis for some value of $z > z_A$. It then enters the fourth quadrant, and both $\mathscr{P}$ and $-\mathscr{Q}$ increase without bound, much as in the case of area I. The curve 1 therefore does not represent an eigensolution.

Curve 3 shows another possibility, that would occur for values of $\varkappa$ and ω slightly different from those of curve 1. In this case, the phase point crosses into the second quadrant, and $-\mathscr{P}$ and $\mathscr{Q}$ increase without bound. Finally, curve 2 would occur for intermediate values of $\varkappa$ and ω: here the phase point approaches the origin as z increases. Moreover, since $\vartheta_a = \pi/2$, the path has a vertical tangent at the origin. In this case, $\mathscr{P}$ and $\mathscr{Q}$ remain finite for all values of z and curve 2 therefore represents an eigensolution.

This discussion indicates that eigensolutions are rare phenomena—that they occur as limiting cases of the more frequent types of solutions represented by curves 1 and 3. It is entirely consistent with the existence of the "natural boundary condition" discussed in Section 79.

Figure 66 does not, of course, represent the only eigensolution of area II. Figure 67 shows a sequence of eigensolutions. In the case $n = 0$, the path does not cross the $\mathscr{Q}$-axis, but z_A occurs in the first quadrant. In the case $n = 1$, the path crosses the $\mathscr{Q}$-axis once, and z_A occurs in the third quadrant. (The reader should convince himself that, if z_A occurs in the second quadrant, no eigensolution is possible.) The case $n = 2$ is that of Fig. 66, and for $n = 3$, the path crosses the $\mathscr{Q}$-axis three times.

The integer n, as defined in this way, is clearly the ordinal number already encountered in earlier sections. The case $n = 0$ is that which has earlier been called the Lamb wave, for which $\mathscr{P}$ has no root or node, other than $z = \infty$. The others are acoustic modes, for which $\mathscr{P}$ has n roots.

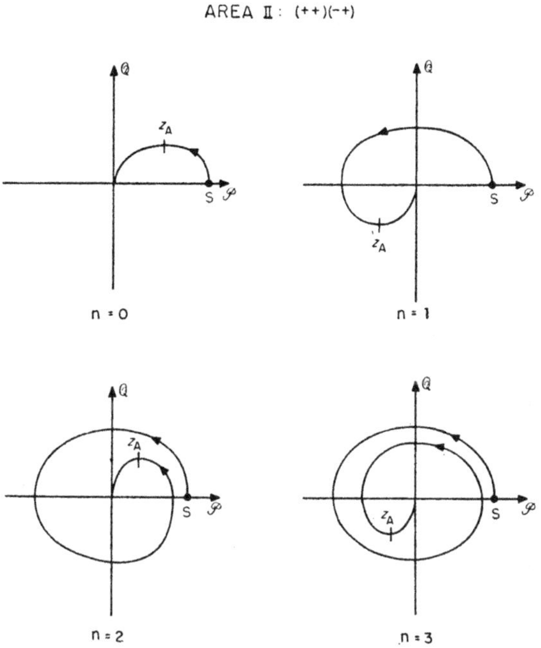

FIG. 67. Sketches of phase paths for the eigensolutions $n = 0, 1, 2, 3$, for AREA II.

AREA IV[b]

In this area, both z_A and z_B are greater than S, and therefore in the physically significant part of the z-axis. They divide the z-axis into three segments with the signature $(+-)\ (++)\ (-+)$. Figure 68 shows the general nature of the eigensolutions for $n = 0$ and $n = 1$. By referring to the theorems of Chapter XIV, the reader should be able to convince himself of the qualitative correctness of these diagrams, and also construct the figures for larger values of n.

AREA III

In this area, only B changes sign, at $z = z_B$, and the signature is $(--)\ (-+)$. Figure 69 shows the eigensolutions for $n = -1$ and -2; negative values of the ordinal number are used to indicate that, now, the phase path crosses the $\mathscr{Q}$-axis in the clockwise or negative direction. The reader should convince himself that the theorems of Chapter XIV are incompatible with the occurrence of $n \geqslant 0$ in this area of the diagram.

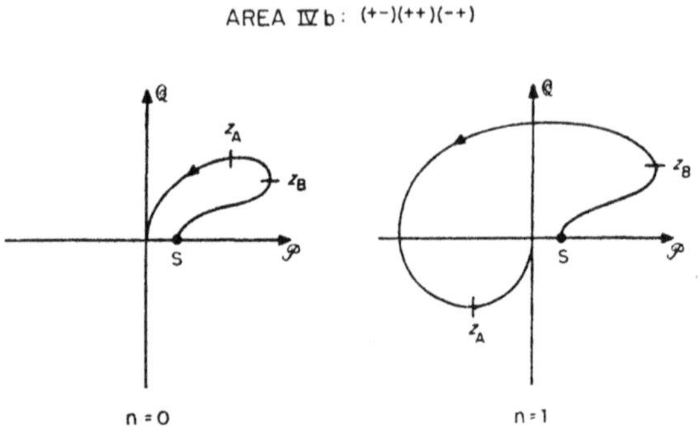

Fig. 68. Sketches of phase paths for the eigensolutions $n = 0, 1$, for Area IV^b.

Because of the location of area III in the $(\varkappa, \omega)$ diagram, it should be evident that these eigensolutions are internal gravity waves. The association of a $(++)$ segment with acoustic waves and a $(--)$ segment with gravity waves, will be found to be characteristic.

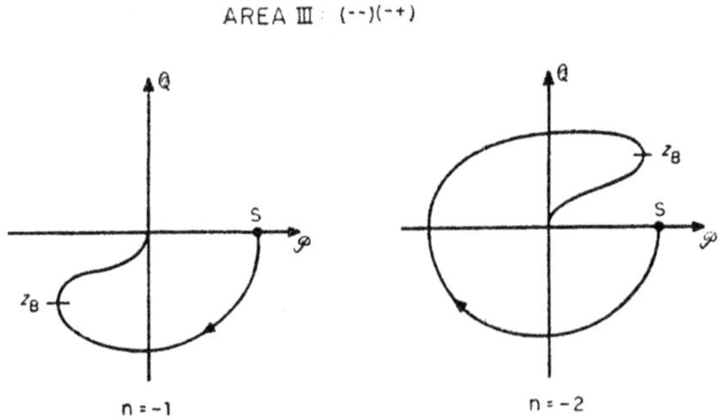

Fig. 69. Sketches of phase paths for the eigensolutions $n = -1, -2$, for Area III.

Area IV^a

Both A and B change sign when $(\varkappa, \omega)$ is a point of this area. From the signature $(+-)\ (--)\ (-+)$, it is possible to deduce that the eigensolutions have phase paths of the kind shown on Fig. 70.

87. The modal curves and the comparison theorem

The phase paths sketched in the preceding section have one important characteristic in common, and this is an accurately quantitative one. It can most readily be formulated in terms of the polar coordinate, ϑ, of the phase diagram. At $z = S$, this angle is zero in every case: the boundary condition at the rigid lower boundary insures this.

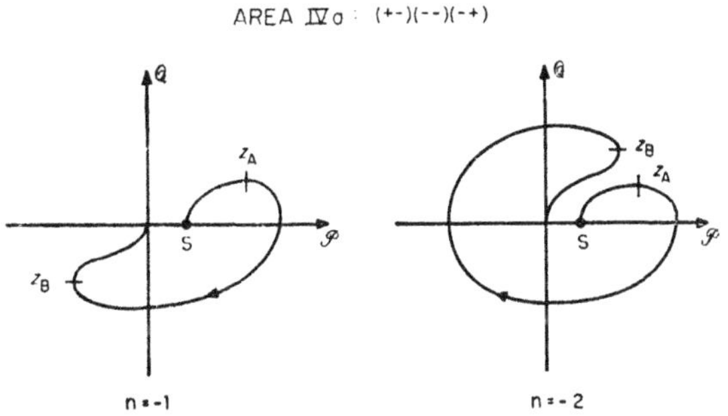

FIG. 70. Sketches of phase paths for the eigensolutions $n = -1, -2$, for AREA IV[a].

As z increases, ϑ may either increase or decrease, but approaches a limit as $z \to \infty$. This limit is easily read from the figures of the last section: it is $(n + \frac{1}{2})\pi$, where n is the ordinal number. This result is valid for both positive and negative ordinals, and therefore justifies the use of negative integers more fully than has hitherto been possible.

Using the notation of Section 84, this result may be written

$$\vartheta(\infty, \mu, \nu) = (n + \tfrac{1}{2})\pi, \tag{1}$$

where

$$\mu = \omega^2/\varkappa^2, \qquad \nu = \omega^2. \tag{2}$$

This is an equation connecting $(\varkappa, \omega)$ and the ordinal number n, and is therefore the equation of the modal curves in the diagnostic diagram.

The theorem proven in Section 84 states that $\vartheta(\infty, \mu, \nu)$ is a monotonically-increasing function of both μ and ν. It remains to interpret this result in terms of the modal curves. Let μ, ν and $\mu + \delta\mu$, $\nu + \delta\nu$ be values corresponding to the same value of n: then Eq. (87–1) yields

$$\frac{\partial\vartheta}{\partial\mu}\,\delta\mu + \frac{\partial\vartheta}{\partial\nu}\,\delta\nu = 0 \tag{3}$$

and hence

$$\delta\mu/\delta\nu = -\left(\frac{\partial\vartheta}{\partial\nu}\bigg/\frac{\partial\vartheta}{\partial\mu}\right). \tag{4}$$

Since both derivatives are positive,

$$\delta\mu/\delta\nu < 0.$$

Using Eq. (87–2) this inequality may be rearranged as

$$0 < \frac{\delta\omega}{\delta\varkappa} < \frac{\omega}{\varkappa}. \tag{5}$$

This is illustrated on Fig. 71; in other words, it states that the slope

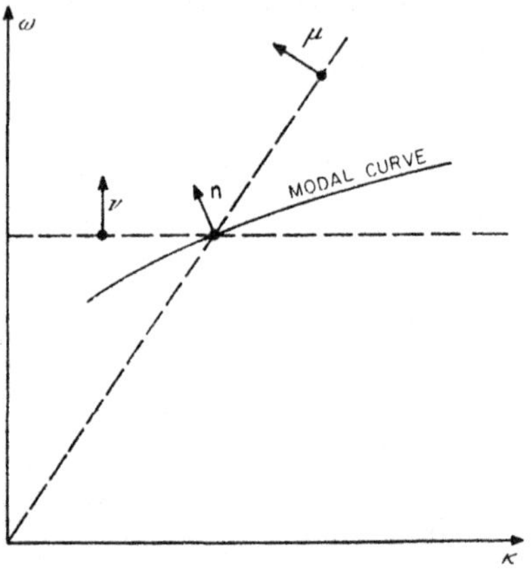

FIG. 71. Illustrating the application of Sturm's comparison theorem to the modal curves in the diagnostic diagram.

of a modal curve in the diagram is always less than the slope of the line drawn to the origin, but greater than zero.

The theorem can also be used to determine the relative positions of modal curves with different ordinal numbers. Let $n' > n''$ be two integers; then the curve for n' is above and to the left of that for n''.

These two results, combined with those of the previous section, would make it possible to sketch the modal curves. Actually, this has already been done in Section 80, using the W-K-B approximation. The reader should convince himself that Fig. 55 is consistent with the results of this section.

Equation (87–5) may also be given another interpretation. In Section 40, the quantity

$$c_T = \omega/\varkappa \tag{6}$$

was defined as the trace velocity of the waves. Its interpretation follows from the general form of the eigensolutions for plane level surfaces; this is

$$\begin{aligned} &\mathscr{P}(z) \exp [i(\alpha x + \beta y - \omega t)], \\ &\mathscr{Q}(z) \exp [i(\alpha x + \beta y - \omega t)]. \end{aligned} \tag{7}$$

Since $\varkappa = (\alpha^2 + \beta^2)^{\frac{1}{2}}$, c_T is the phase speed associated with the exponential factors of Eq. (87–7). If the ordinal number is held constant, Eq. (87–1) makes ω a function of $\varkappa$. One may therefore also associate a group velocity to the exponential factors, using the same arguments as in Sections 44 and 64. It is readily shown that this is

$$c_{TG} = (\partial\omega/\partial\varkappa)_{n=\text{const.}} \tag{8}$$

In other words, the trace group velocity is the slope of the modal curve. Equation (87–5) then becomes

$$0 < c_{TG} < c_T. \tag{9}$$

88. An atmosphere with a single temperature minimum

The stratification of the Earth's atmosphere has been described in Section 29. Figure 13 shows that its temperature has two minima and one maximum. The frequency, N, has two minima and two maxima. These various extrema create many possible types of eigensolution, each characterized by a different signature. The number of these is somewhat reduced if, as was assumed in preparing Fig. 13, there are equalities among the temperature extrema, but remains very large, even then.

In order to illustrate the complexities that arise, without the necessity of considering an unduly great number of special cases, this section will be devoted to the study of an atmosphere whose temperature has a single minimum, as shown on Fig. 72. The stability frequency, N, was calculated from Eq. (25–2), the temperature gradient being obtained graphically. It shows a single minimum at an altitude somewhat less than that of the temperature minimum. The region below about 15 km, in this hypothetical case, will have properties similar to the Earth's troposphere. There is also a maximum value of N at about 25 km; this stratum will correspond to the Earth's stratosphere. The thermosphere corresponds to the region of increasing temperature and decreasing stability.

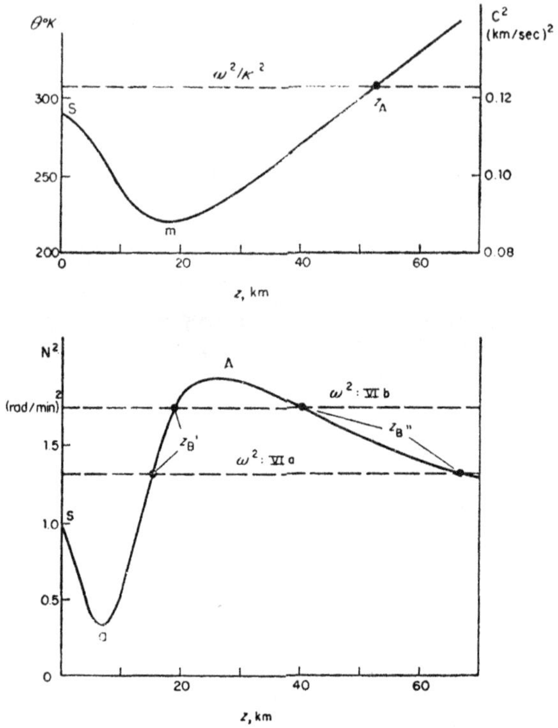

FIG. 72. Illustrating the change of signature when ω^2 varies and $\omega^2/\varkappa^2$ is constant.

The complexity of the problem arises from the fact that the coefficients of the canonic equations may each have several roots. Thus, if

$\omega/\varkappa > c_S$,	A will have one root;
$c_S > \omega/\varkappa > c_m$,	A will have two roots;
$c_m > \omega/\varkappa$,	A will have no root;

while, if

$\omega > N_A$,	B will have no root;
$N_A > \omega > N_S$,	B will have two roots;
$N_S > \omega > N_a$,	B will have three roots;
$N_a > \omega$,	B will have one root.

Figure 73 shows the diagnostic diagram divided into twelve areas by straight lines, each of the areas being characterized by a definite number of roots for the coefficients A and B. For example, those $(\varkappa, \omega)$ values that lie in area VI lead to one root for A and two roots for B; this is illustrated by the horizontal lines drawn on Fig. 72.

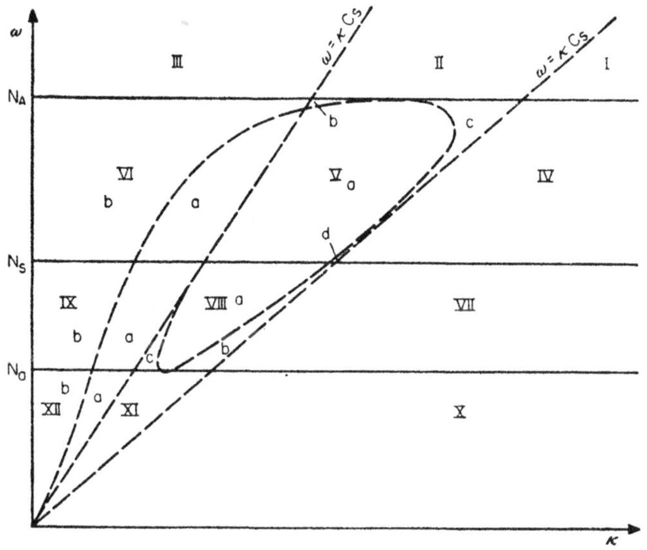

FIG. 73. Subdivision of the diagnostic diagram for an atmosphere with a single temperature minimum. Signatures as shown in the table.

Throughout area VI, the roots of A and B will divide the z-axis into four segments; but the signatures of the segments depend on the order of these roots, as well as on their number. Consequently, area VI is divided into two parts by the curved line whose equation is

$$z_A = z_B'',$$

the notation being that illustrated on Fig. 72. Consequently, in

area VI^a, $S < z_B' < z_A < z_B''$: $(++)$ $(+-)$ $(--)$ $(-+)$;

while in

area VI^b, $S < z_B' < z_B'' < z_A$: $(++)$ $(+-)$ $(++)$ $(-+)$.

These two cases are illustrated on Fig. 72. The areas V, VIII, IX and XII are also subdivided in a similar fashion.

When the possible segmentation of the z-axis is carefully examined, it is found that the comparatively simple temperature graph of Fig. 72 leads to twenty different signatures. These are listed below, the grouping being based on a fairly obvious system. Again, all of them end with the characteristic thermosphere signature, (−+).

Signature						Area
(−+)						I
(−+)	(++)	(−+)				II
(++)	(−+)					III
(−+)	(++)	(+−)	(++)	(−+)		V^b
(++)	(+−)	(++)	(−+)			VI^b
(+−)	(++)	(+−)	(++)	(−+)		IX^b
(+−)	(++)	(−+)				XII^b
(−+)	(−−)	(−+)				IV
(−+)	(−−)	(+−)	(−−)	(−+)		V^d
(−−)	(−+)	(−−)	(−+)			VII
(−−)	(−+)	(−−)	(+−)	(−−)	(−+)	VIII^b
(−−)	(−+)					X
(−−)	(+−)	(−−)	(−+)			XI
(+−)	(−−)	(−+)				XII^a
(−+)	(++)	(+−)	(−−)	(−+)		V^a
(−+)	(++)	(−+)	(−−)	(−+)		V^c
(++)	(+−)	(−−)	(−+)			VI^a
(−−)	(−+)	(++)	(+−)	(−−)	(−+)	VIII^a
(−−)	(+−)	(++)	(+−)	(−−)	(−+)	VIII^c
(+−)	(++)	(+−)	(−−)	(−+)		IX^a

Five of these signatures have already been studied in the last section; these are areas I, III, X, XII^a and XII^b. As examples of the remaining fifteen signatures, it is most instructive to consider the areas VI^a and VI^b, since they illustrate certain general principles without undue complication.

Area VI^a

Figure 74 illustrates two possible phase paths of this area. The path shown on the left of this figure will be described first. The segment (++) extends from the ground level to z_B'; it is a spiral described in the positive sense, crossing the $\mathscr{Q}$-axis three times. The point z_B' therefore occurs in the third quadrant, and the path has a vertical tangent here. The (+−) segment extends from z_B' to z_A, and ends with a horizontal tangent; in this segment, the radius vector has increased appreciably. It is followed by the (−−) segment, which is described in the clockwise or negative sense. The radius vector remains large throughout this segment, which ends in the first quadrant at z_B''. In the final thermosphere segment, the path approaches the origin along the $\mathcal{Q}$-axis.

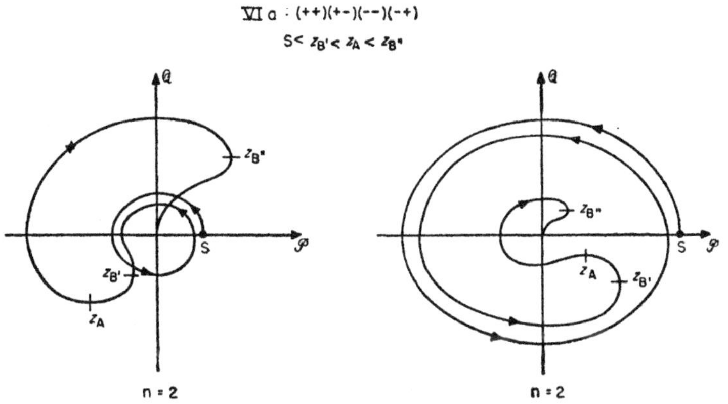

FIG. 74. Sketches of two possible eigensolutions for $n = 2$ and the signature $(++)(+-)(--)(-+)$, AREA VI[a].

The phase path on the right of Fig. 74 differs from that on the left in two major respects. First, it has been drawn to a larger scale; the initial $(++)$ segment is intended to be nearly similar in the two cases; they are intended to represent $(\varkappa, \omega)$ points that are not widely separated. Because of this difference in the parameter values, the $(++)$ segment now ends in the fourth quadrant. Consequently the ensuing $(+-)$ segment results in a marked decrease of the radius vector (cf. Fig. 59). The radius vector then remains small throughout the $(--)$ and $(-+)$ segments.

This illustrates a general principle: when the signature contains two oscillatory segments separated by a non-oscillatory one, small changes in $\varkappa$ and ω may produce large quantitative changes in the phase path. In particular, the right-hand part of Fig. 74 represents an oscillation that has a relatively small amplitude at low altitudes, and a large amplitude in the $(--)$ channel. The left-hand part represents a large amplitude in the low $(++)$ channel and a relatively small amplitude aloft.

There is another difference in the two phase paths: the one on the left crosses the $\mathcal{Q}$-axis three times in the positive sense and once in the negative; the one on the right crosses it four times positively and twice negatively. The polar angle, ϑ, thus attains different maximum values (between $z = z_B{}'$ and $z = z_A$), in the two cases. However, as $z \to \infty$, $\vartheta \to 5\pi/2$ in both cases; hence one will assign the ordinal number $n = 2$ to both, on the basis of Eq. (68–1).

AREA VI^{b}

The signature of this area differs from that just described, primarily in that both oscillatory segments are $(++)$. Consequently the angular velocity of the phase point is positive in all segments, whereas

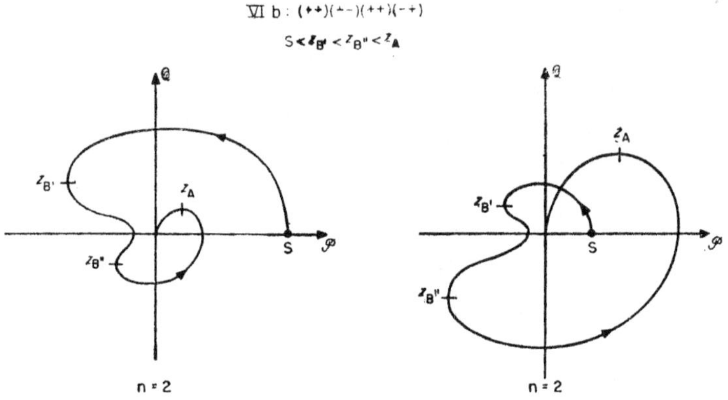

FIG. 75. Sketches of two possible eigensolutions for $n = 2$ and the signature $(++)(+-)(++)(-+)$, AREA VI^{b}.

in the previous case, it changed sign. All roots of $\mathscr{P}$ are thus positive, and the original definition of the ordinal number suffices to show that both paths illustrated on Fig. 75 correspond to $n = 2$. The paths have again been sketched to different scales, and again illustrate the large quantitative changes that can result from small changes in the values of $\varkappa$ and ω.

The study of the remaining thirteen signatures reveals no further general principles and will therefore be omitted. In concluding these chapters on the residual equations, it may be remarked that the canonic form, Eq. (82–8), seems especially well-adapted for numerical solution, and that the qualitative methods used here can therefore be supplemented by quantitative results whenever necessary.

89. The modal equation for an ocean of constant depth

The qualitative methods used above can also be applied to the eigensolutions and modal curves of an ocean of constant depth. The general procedures based on the roots of $A(z)$ and $B(z)$, and the signatures of the solutions, all remain unchanged. The single new feature is the boundary condition at the free surface, Eq. (54–1):

$$cP + gQ = 0 \qquad \text{at the free surface.} \tag{1}$$

This must be expressed in terms of the fields $\mathscr{P}$ and $\mathscr{Q}$; using the equations of Section 82, it is found that

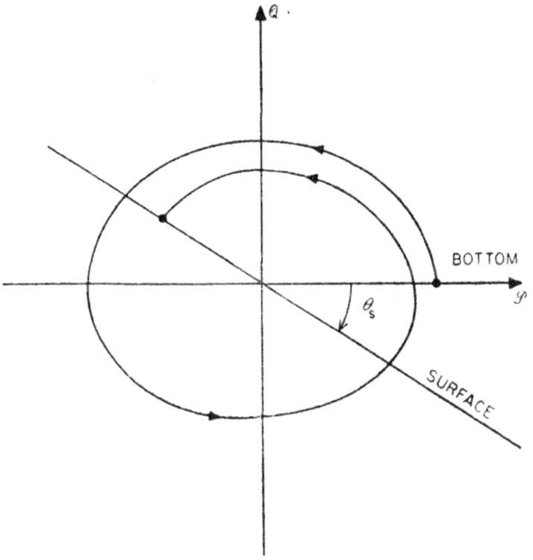

FIG. 76. Illustrating the boundary condition at the free surface of an ocean or lake.

$$\mathscr{Q} + \mathscr{P} \exp(-2\delta_S) = 0, \tag{2}$$

where

$$\delta_S = g \int_{bottom}^{surface} \frac{1}{c^2}\, dz. \tag{3}$$

If the velocity of sound is constant

$$\delta_S = gD/c^2;$$

it is therefore clear that δ_S will be small for all of the Earth's oceans.

The boundary condition at the rigid bottom remains

$$\mathscr{Q} = 0 \quad \text{at the bottom.} \tag{4}$$

In the phase diagram, the bottom is therefore represented by the $\mathscr{P}$-axis, while the free surface is represented by the straight line determined by Eq. (89–2) and shown on Fig. 76. This passes through the origin, and makes the angle $-\vartheta_S$ with the $\mathscr{P}$-axis, where

$$\tan \vartheta_S = \exp(-2\delta_S). \tag{5}$$

It is seen that

$$0 \leqslant \vartheta_S \leqslant \pi/4, \tag{6}$$

and that, for shallow oceans, ϑ_S is nearly $\pi/4$.

The phase path must start from the bottom or $\mathscr{P}$-axis, cross the $\mathscr{Q}$-axis n times, and end on the surface lines. The equation of the modal lines is therefore

$$\vartheta(\mu, \nu) = n\pi - \vartheta_S, \tag{7}$$

provided the polar angle is measured from the $\mathscr{P}$-axis.

It follows that all of the general conclusions concerning the slope and relative position of the modal curves, as derived in Section 87 for the atmosphere, apply equally to the oceans.

CHAPTER XVI

THE WAVE EQUATION FOR A SPHERICAL SURFACE

90. Introduction

THE mathematical problems presented by spherical level surfaces without rotation are only slightly more complex than those raised by plane level surfaces. As shown in Section 36, the variables separate and the expressions for the fields factorize:

$$P = P_1(\lambda, \mu)P_2(\chi), \quad \text{etc.,} \tag{1}$$

and P_2, Q_2 satisfy the residual equations, Eq. (36–12), in practically unaltered form. The factor P_1 must satisfy the wave equation for a spherical surface (Eq. 36–23), which is

$$\cosh^2 \mu \left(\frac{\partial^2 P_1}{\partial \lambda^2} + \frac{\partial^2 P_1}{\partial \mu^2} \right) + \varkappa^2 r_S{}^2 P_1 = 0, \tag{2}$$

r_S being the radius of the planet's surface. The close resemblance between this and the two-dimensional wave equation (Eq. 36–11), has already been noted.

However, there is one difference between the two equations. If the fluid has no lateral boundaries, and the level surfaces are plane, there is an eigensolution for every value of $\varkappa$. This is not true of Eq. (90–2): it has an eigensolution only for a discrete set of values of $\varkappa r_S$. This is analogous to the plane case of a tank with lateral boundaries. The reason is clear: the area of an unbounded plane is infinite, while that of a whole sphere is finite. It is therefore not surprising that the horizontal ordinal numbers of a tank have their analogues when the fluid covers the whole of a spherical planet.

After the calculations for this atmospheric case have been completed, the remainder of this chapter will be devoted to the study of finite bodies of water on a spherical surface. As in Section 33, it will be noted that a general bottom topography presents mathematical difficulties that preclude a brief discussion. The somewhat artificial case of an ocean with level bottom and vertical walls presents fewer difficulties, in the sense that the residual equations and Eq. (90–2) still apply. However, the solution for arbitrary shapes is still too difficult to be attempted without numerical computation. Only if the boundary is a spherical rectangle, with walls along meridians and circles of latitude, can the solution be discussed in some detail.

91. The Legendre functions

Separation of variables. Equation (90–2) has solutions of the form

$$P_1(\lambda, \mu) = Z(\mu) \exp(ism\lambda), \tag{1}$$

where

$$s = \pm 1, \tag{2}$$

and m is a positive constant. It will ultimately be found that the difference between the cases $s = +1$ and $s = -1$ is great enough to justify this notation, although the factor s could be omitted if negative values m were not excluded. If the time dependence of the eigensolutions (cf. Section 34) is included, they have the factor

$$\exp[i(sm\lambda - \omega t)]$$

This represents a wave travelling in the west-east direction with the speed

$$sm/\omega \quad \text{radians per second.}$$

If $s = +1$, the wave travels from west to east—in the positive direction; if $s = -1$, it travels from east to west. Thus, s is the sign of the speed. Without rotation, the sign of the speed is of little importance; symmetry considerations can be used to obtain solutions for $s = -1$ from those for $s = +1$. However, rotation about the polar axis of the co-ordinates will destroy this symmetry, and create important differences between positive and negative waves; this will appear only in the next chapter, but the distinction will be introduced here, to facilitate later comparison.

Since (μ, λ) and $(\mu, \lambda + 2\pi)$ are the same points on the sphere, the function P_1 defined by Eq. (91–1) will be single-valued only if

$$m = 0, 1, 2, 3, \cdots. \tag{3}$$

This is an essential requirement if the fluid covers the whole sphere like an atmosphere. If it does not, the restriction of m to integral values must be replaced by another (see Section 93). The integer m will be called the longitude ordinal number, and is analogous to one of the numbers (k, m) encountered in Section 59. In quantum theory, sm is called the magnetic quantum number.

At the poles, $\mu = \pm\infty$, and all values of λ represent the same point on the sphere. The requirement of single-valuedness therefore imposes the "natural boundary condition."

$$\begin{aligned} Z(\pm\infty) &= 0 \qquad \text{if } m \neq 0, \\ &= \text{finite, if } m = 0. \end{aligned} \tag{4}$$

The behaviour of the velocity components at the poles will require further consideration later (cf. Eq. 91–29).

The Legendre equation. If Eq. (91–1) is substituted into Eq. (90–2), it is found that

$$\frac{\mathrm{d}^2 Z}{\mathrm{d}\mu^2} = (m^2 - \varkappa^2 r_S{}^2 \operatorname{sech}^2 \mu) Z. \tag{5}$$

This equation is of such a form that the qualitative methods of Section 72 may be applied to it at once. The convexity of the (μ, Z) graph is the parenthesis of Eq. (91–5), and the graph will therefore oscillate when this parenthesis is negative (cf. Fig. 42). This will occur when $m \neq 0$ and

$$\cosh \mu < \varkappa r_S/m.$$

Outside the zone of latitudes defined by this inequality, the function behaves in a pseudo-exponential fashion, and must approach zero for $\mu = \pm\infty$ (cf. Eq. 90–4). If $m = 0$, then Z oscillates over the whole range of μ.

The Legendre functions. Throughout this and the following chapter, the notation

$$\begin{aligned} \tau &= \tanh \mu, \\ \sigma &= \operatorname{sech} \mu = (1 - \tau^2)^{\frac{1}{2}}, \end{aligned} \tag{6}$$

will be convenient. Very often, τ will be used instead of μ as independent variable:

$$\frac{\mathrm{d}}{\mathrm{d}\mu} = \sigma^2 \frac{\mathrm{d}}{\mathrm{d}\tau} = (1 - \tau^2) \frac{\mathrm{d}}{\mathrm{d}\tau}. \tag{7}$$

On introducing τ as variable in Eq. (91–5), and defining

$$L = \varkappa r_S, \tag{8}$$

it becomes

$$\frac{\mathrm{d}}{\mathrm{d}\tau}\left(\sigma^2 \frac{\mathrm{d}Z}{\mathrm{d}\tau}\right) + (L^2 - m^2/\sigma^2)Z = 0, \tag{9}$$

which is known as Legendre's equation.[1,2,3]

In general, it has no solution that satisfies Eq. (91–4); however, if

$$L^2 = l(l+1), \qquad l = m, m+1, \cdots m+k \cdots, \tag{10}$$

then it does have such a solution. This is the Legendre function,

$$Z = P_l^{\,m}(\tau), \tag{11}$$

defined by the formulae

$$P_l^{\,0}(\tau) = P_l(\tau) = \frac{1}{2^l l!}\left(\frac{\mathrm{d}}{\mathrm{d}\tau}\right)^l (\tau^2 - 1)^l, \tag{12}$$

$$P_l^{\,m}(\tau) = (-1)^m (1 - \tau^2)^{m/2} \left(\frac{\mathrm{d}}{\mathrm{d}\tau}\right)^m P_l(\tau). \tag{13}$$

The first few Legendre functions are given in Table 3. It will be

TABLE 3.

Legendre functions

$P_0^{\,0} = 1$	$P_2^{\,2} = 3\sigma^2$
$P_1^{\,0} = \tau$	$P_3^{\,0} = (5\tau^3 - \tau)/2$
$P_1^{\,1} = -\sigma$	$P_3^{\,1} = -3\sigma(5\tau^2 - 1)/2$
	$P_3^{\,2} = 15\sigma^2\tau$
$P_2^{\,0} = (3\tau^2 - 1)/2$	$P_3^{\,3} = -15\sigma^3$
$P_2^{\,1} = -3\sigma\tau$	

noted that $P_l(\tau)$ is a polynomial, of degree l, in τ; its roots are all real and occur in the range $|\tau| < 1$. Similarly,

$$\left(\frac{\mathrm{d}}{\mathrm{d}\tau}\right)^m P_l(\tau)$$

is a polynomial in τ, of degree $k = l - m$; all k of its roots occur in $|\tau| < 1$. In addition, the factor

$$\sigma^m = (1 - \tau^2)^{m/2}$$

has roots of degree $m/2$ at $\tau = \pm 1$. In various senses, therefore, k and l are again the number of roots of Z. If the multiple polar roots are excluded, Z has exactly $l - m = k$ roots.

In quantum theory, l is called the azimuthal quantum number; here it will be called the azimuthal ordinal number of the eigensolution. Similarly k will be called the latitude ordinal number.

The Legendre functions satisfy numerous relations, many of which can be obtained from the two fundamental formulae

$$\sigma^2 \frac{\mathrm{d}P_l^m}{\mathrm{d}\tau} + l\tau P_l^m = (l + m)P_{l-1}^m, \tag{14}$$

$$\sigma^2 \frac{\mathrm{d}P_{l-1}^m}{\mathrm{d}\tau} - l\tau P_{l-1}^m = (l - m)P_l^m. \tag{15}$$

For example, these two last equations imply that

$$(2l + 1)\tau P_l^m = (l + 1 - m)P_{l+1}^m + (l + m)P_{l-1}^m. \tag{16}$$

Expressions for the velocities. Equations (36–13, 14, 17) are

$$i\omega U_\epsilon = \frac{c}{r\sigma}\frac{\partial P}{\partial \lambda}, \tag{17}$$

$$i\omega U_\nu = \frac{c}{r\sigma}\frac{\partial P}{\partial \mu}, \tag{18}$$

$$i\omega Q = U_\zeta. \tag{19}$$

When the substitution (see Eq. (91–1))

$$P = Z(\mu)\mathrm{e}^{ism\lambda}P_2(\chi), \tag{20}$$

$$Q = Z(\mu)\mathrm{e}^{ism\lambda}Q_2(\chi), \tag{21}$$

is introduced into these equations, the velocity components are found to have the form

$$U_\epsilon = \frac{c}{\omega r \sigma} P_2(\chi) X(\mu) e^{ism\lambda}, \tag{22}$$

$$U_\nu = \frac{c}{i\omega r \sigma} P_2(\chi) Y(\mu) e^{ism\lambda}, \tag{23}$$

$$U_\zeta = i\omega Q_2(\chi) Z(\mu) e^{ism\lambda}. \tag{24}$$

The functions X, Y and Z are found to be related by the equations

$$X = sm\, Z, \tag{25}$$

$$Y = \frac{dZ}{d\mu}, \tag{26}$$

$$L^2\sigma^2 Z = smX - \frac{dY}{d\mu}. \tag{27}$$

When X and Y are eliminated from these three equations, Eq. (91–9) is recovered.

Using $Z = P_l^m(\tau)$ and Eq. (91–16), it is found that

$$Y = -l\tau P_l^m + (l+m) P_{l-1}^m, \tag{28}$$

a formula that is convenient for computational purposes.

The first few eigensolutions are listed in Table 4.

TABLE 4.

Eigensolutions

$k = 0$,	$l = 0$,	$m = 0$:	$Z = 1$,	$Y = 0$
$k = 1$,	$l = 1$,	$m = 0$:	$Z = \tau$,	$Y = \sigma^2$
$k = 0$,	$l = 1$,	$m = 1$:	$Z = \sigma$,	$Y = -\sigma\tau$
$k = 2$,	$l = 2$,	$m = 0$:	$Z = (3\tau^2 - 1)/2$,	$Y = 3\sigma^2\tau$
$k = 1$,	$l = 2$,	$m = 1$:	$Z = -3\sigma\tau$,	$Y = 3\sigma(2\tau^2 - 1)$
$k = 0$,	$l = 2$,	$m = 2$:	$Z = 3\sigma^2$,	$Y = -6\sigma^2\tau$

It remains to consider whether the velocity fields corresponding to these eigensolutions are all regular at the poles, where $\tau = \pm 1$, $\sigma = 0$. This will certainly be the case if all three components vanish for $\sigma = 0$, and an inspection of this table and Eqs. (91–22, 23, 24) show that this will be the case except when $m = 0$ or 1. If $m = 0$, both U_ϵ and U_ν vanish in every case, while U_ζ is finite, but independent of latitude; this is a satisfactory situation. If $m = 1$, U_ζ vanishes at the poles, but U_ϵ and U_ν do not, and depend on the latitude. It will suffice to consider the case $l = 1$, $m = 1$; then

$$U_\epsilon = \frac{cs}{\omega r} P_2[\cos(s\lambda) + i \sin(s\lambda)],$$

$$U_\nu = \frac{c}{\omega r} P_2[-\sin(s\lambda) + i \cos(s\lambda)].$$

Since λ is essentially the polar co-ordinate in a plane tangent to the sphere at the pole, it follows that this velocity field has no singularity.

The natural boundary conditions. The natural boundary conditions can now be restated more completely: at the poles,

$$\begin{aligned} &\text{if } m = 0: && X/\sigma = Y/\sigma = 0, && Z = \text{finite}; \\ &\text{if } m = 1: && X/\sigma = Y/\sigma = \text{finite}, && Z = 0; \\ &\text{if } m > 1: && X/\sigma = Y/\sigma = Z = 0. && \end{aligned} \tag{29}$$

The phase diagrams. Eliminating χ by Eq. (91–25), Eqs. (91–26) and (91–27) become

$$\frac{dZ}{d\mu} = Y, \tag{29}$$

$$\frac{dY}{d\mu} = -(L^2\sigma^2 - m^2)Z, \tag{30}$$

which have the canonic form of Section 82; the identification is

$$z = \mu$$

$$Y = \mathscr{P}, \quad Z = \mathscr{Q}, \tag{31}$$

$$A = 1, \qquad B = L^2 \operatorname{sech}^2\mu - m^2.$$

Phase diagrams, using Y and Z as co-ordinates, may therefore be plotted—Fig. 77. Since explicit expressions for these variables are available, the graphs may be made quantitative in this case. Equation (91–31) shows that their signatures are $(++)$ if $m = 0$, and $(+-)$ $(++)$ $(+-)$ if $m \neq 0$. It is to be noted that Y and Z are independent of the sign, s, of the speed, but that $X = smZ$.

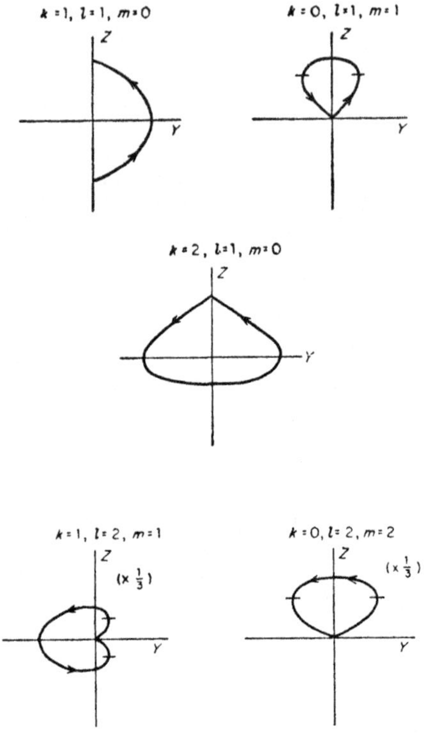

FIG. 77. Eigensolutions of the Legendre equation.

92. Segmental ocean

In accordance with the plan outlined in Section 90, the next problem to be examined is that of an ocean bounded by vertical shores. The simplest case is that of a segment, extending from pole to pole, and bounded by great semi-circles. These may be chosen as the meridians $\lambda = 0$ and $\lambda = \Lambda > 0$; they bring with them the boundary conditions

$$U_\epsilon = 0 \quad \text{at} \quad \lambda = 0 \quad \text{and} \quad \lambda = \Lambda. \tag{1}$$

Reference to Eq. (91–17) shows that this condition will be satisfied if

$$P = Z(\mu)\cos(fm\lambda)P_2(\chi),\qquad Q = Z(\mu)\cos(fm\lambda)Q_2(\chi) \tag{2}$$

and

$$f = \pi/\Lambda, \qquad m = 0, 1, 2\ldots. \tag{3}$$

Then Eqs. (91–17, 18, 19) yield

$$U_\epsilon = -\frac{c}{i\omega r\sigma}X(\mu)\sin(fm\lambda)P_2(\chi),$$

$$U_\nu = \frac{c}{i\omega r\sigma}Y(\mu)\cos(fm\lambda)P_2(\chi), \tag{4}$$

$$U_\zeta = i\omega Z(\mu)\cos(fm\lambda)Q_2(\chi),$$

where

$$X = fmZ, \tag{5}$$

$$Y = \mathrm{d}Z/\mathrm{d}\mu, \tag{6}$$

$$L^2\sigma^2 Z = fmX - \mathrm{d}Y/\mathrm{d}\mu. \tag{7}$$

When X and Y are eliminated from these three equations, the result is

$$\frac{\mathrm{d}}{\mathrm{d}\tau}\left(\sigma^2\frac{\mathrm{d}Z}{\mathrm{d}\tau}\right) + (L^2 - m^2f^2/\sigma^2)Z = 0, \tag{8}$$

which differs from Legendre's equation only in that mf replaces m. This is not entirely trivial, since mf is no longer an integer.

The eigenvalues of this equation may again be written

$$L^2 = l(l+1), \tag{9}$$

but l is not necessarily an integer. Instead,

$$l = mf + k, \quad k = 0, 1, 2\ldots. \tag{10}$$

Fig. 78 is a graph of L as a function of Λ, for various values of m and k. The eigensolutions have a slightly different form when k is even than when k is odd.

When k is even

$$Z = A\sigma^{mf}\left[1 - \frac{k(2mf+k+1)}{2!}\tau^2 + \frac{k(k-2)(2mf+k+1)(2mf+k+3)}{4!}\tau^4 + \ldots\right] \tag{11}$$

and when k is odd,

$$Z = A\sigma^{mf}\left[\tau - \frac{(k-1)(2mf+k+2)}{3!}\tau^3 + \frac{(k-1)(k-3)(2mf+k+2)(2mf+k+4)}{5!}\tau^5 + \ldots\right] \quad (12)$$

The constants A are conveniently chosen so that these formulae reduce to

$$Z = P_l^m(\tau)$$

when $f = 1$.

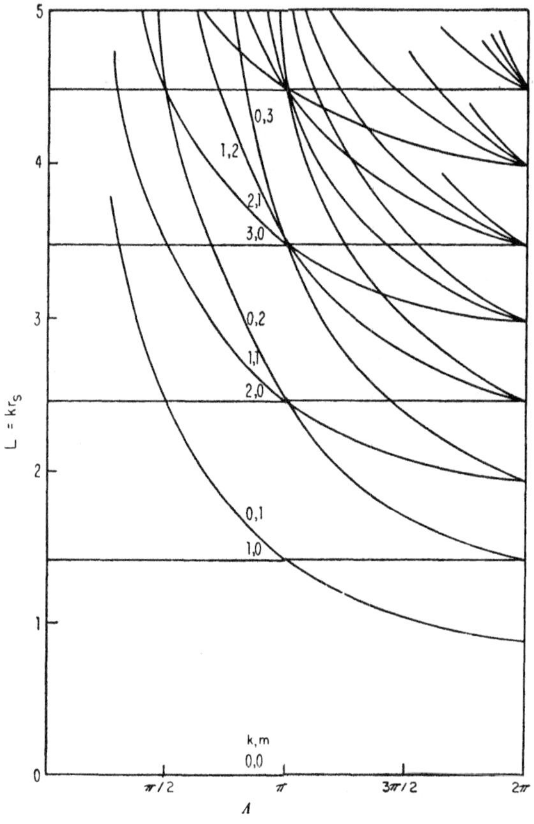

FIG. 78. Dependence of the eigenvalue L on the angular width, Λ of a segmental ocean.

The first few eigensolutions are given in Table 5.

TABLE 5

Eigensolutions

$k = 0,$	$l = 0,$	$m = 0:$	$Z = 0,$	$Y = 0$
$k = 1,$	$l = 1,$	$m = 0:$	$Z = \tau,$	$Y = \sigma^2$
$k = 0,$	$l = f,$	$m = 1:$	$Z = \sigma^f,$	$Y = -f\sigma^f\tau$
$k = 2,$	$l = 2,$	$m = 0:$	$Z = (3\tau^2 - 1)/2$	$Y = 3\sigma^2\tau$
$k = 1,$	$l = f + 1,$	$m = 1:$	$Z = -3\sigma^f\tau,$	$Y = 3\sigma^f[(1 + f)\tau^2 - 1]$
$k = 0,$	$l = 2f,$	$m = 2:$	$Z = 3\sigma^{2f},$	$Y = -6f\sigma^{2f}\tau$

References for Chapter XVI

1. W. E. BYERLY, *Fourier's Series and Spherical Harmonics*, Ginn & Co., New York, 1893.
2. T. M. MACROBERT, *Spherical Harmonics*, Dover, New York, 1948.
3. W. MAGNUS and F. OBERHETTINGER, *Special Functions of Mathematical Physics*, Chelsea, New York, 1949.

CHAPTER XVII

SPHERICAL LEVEL SURFACES WITH ROTATION

93. Introduction

THE mathematical difficulties connected with the combination of Coriolis force and spherical level surfaces have been discussed in Section 38. It was shown there that the most serious difficulty—the inseparability of the equations—can be removed by neglecting the horizontal component of the angular velocity of the planet. The justification of this neglect is another matter, but it has become traditional to accept the consequences of this approximation.

When this tradition is followed, the equations separate into the residual equations (which have already been discussed in earlier chapters) and an equation that involves only λ and μ. This is a generalization of the wave equation on a spherical surface, and was first derived by Laplace in connection with the tides in an ocean of constant depth. It may therefore be called the tidal equation although it is relevant to a much wider class of problems.

Despite the number of papers that have been devoted to this equation, its theory is still in a quite unsatisfactory state. It leads to a generalization of Legendre's equation (Eq. 91–9), whose solutions are not polynomials but can only be expressed in terms of infinite series. Moreover, even the calculation of the coefficients of these series is a problem of some difficulty.

It is certain that a determined effort would yield solutions to all problems connected with the tidal equation, but such completeness will not be attempted here. There are a number of points at which progress is easy, and new results are obtainable without difficulty. Only these matters will be discussed in this chapter.

94. The first-order tidal equations

In Section 38, the Laplace tidal equation was derived in a form that is analogous to the wave equation on a sphere. This is not the most useful form of the equation, and the first task will be to give a second derivation, that yields more flexible equations. The form obtained in the earlier section was

$$\frac{\partial}{\partial \lambda}\left\{\frac{1}{a^2\tau^2 - 1}\left[\frac{\partial P_1}{\partial \lambda} + ia\tau\frac{\partial P_1}{\partial \mu}\right]\right\} + \frac{\partial}{\partial \mu}\left\{\frac{1}{a^2\tau^2 - 1}\left[\frac{\partial P_1}{\partial \mu} - ia\tau\frac{\partial P_1}{\partial \lambda}\right]\right\} - \sigma^2 L^2 P_1 = 0, \quad (1)$$

where

$$a = \Omega/\omega \quad \text{and} \quad L = \varkappa r_S. \quad (2)$$

The substitution

$$P_1 = Z(\mu) \exp(ism\lambda)$$

(cf. Eq. (91–1)) reduces this to the ordinary differential equation.

$$\left(\frac{\mathrm{d}}{\mathrm{d}\mu} - sma\tau\right)\frac{1}{1 - a^2\tau^2}\left(\frac{\mathrm{d}Z}{\mathrm{d}\mu} + sma\tau Z\right) + (L^2\sigma^2 - m^2)Z = 0. \quad (3)$$

The appearance of the factor $1 - a^2\tau^2$ in the denominator makes this an awkward equation to study. It can be given other forms, which will now be derived.

After making the traditional approximation, Eqs. (38–3 . . . 7) are

$$i\omega U_\epsilon + \Omega\tau U_\nu = \frac{c}{r\sigma}\frac{\partial P}{\partial \lambda}, \quad (4)$$

$$i\omega U_\nu - \Omega\tau U_\epsilon = \frac{c}{r\sigma}\frac{\partial P}{\partial \mu}, \quad (5)$$

$$i\omega U_\zeta = c\left(\frac{\partial P}{\partial r} + \Gamma P\right) - N^2 Q, \quad (6)$$

$$i\omega P = \frac{c}{r\sigma}\left[\frac{\partial U_\epsilon}{\partial \lambda} + \frac{1}{\sigma}\frac{\partial}{\partial \mu}(\sigma U_\nu)\right] + c\left[\frac{1}{r^2}\frac{\partial}{\partial r}(r^2 U_\zeta) - \Gamma U_\zeta\right], \quad (7)$$

$$i\omega Q = U_\zeta. \quad (8)$$

If Eqs. (91–22, 23, 24), for the velocities, are substituted into these equations, it is found that P_2 and Q_2 must satisfy the residual equations, while X, Y, Z must satisfy

$$X - a\tau Y = smZ, \tag{9}$$

$$Y - a\tau X = \frac{dZ}{d\mu}, \tag{10}$$

$$smX - \frac{dY}{d\mu} = L^2\sigma^2 Z. \tag{11}$$

It is readily verified that these reduce to Eqs. (91–25, 26, 27) when $a = 0$. Since $a = \Omega/\omega$, the limit of very high frequencies is also $a = 0$, and the effect of rotation will vanish in this limit. This point will be discussed more fully in Section 98.

If X is eliminated between the last three equations, the result is

$$\frac{dZ}{d\mu} + sma\tau Z = (1 - a^2\tau^2)Y, \tag{12}$$

$$\frac{dY}{d\mu} - sma\tau Y = (m^2 - L^2\sigma^2)Z. \tag{13}$$

If Y is also eliminated, Eq. (94–3) is recovered. However, it is much more convenient to work with Eqs. (94–12, 13) than with Eq. (94–3).

The natural boundary conditions for the functions X, Y, Z are the same as before, and are given explicitly as Eq. (91–29).

95. The zonal oscillations

If $m = 0$, the eigensolutions become independent of longitude, and therefore represent zonally-symmetric oscillations. In this case Eqs. (94–12, 13) reduce to

$$\frac{dZ}{d\mu} = (1 - a^2\tau^2)Y = \sigma^2\frac{dZ}{d\tau}, \tag{1}$$

$$\frac{dY}{d\mu} = -L^2\sigma^2 Z = \sigma^2\frac{dY}{d\tau}. \tag{2}$$

The case $L = 0$. If $L = 0$, these equations simplify further, and have the obvious solution

$$Y = 0, \quad Z = 1, \tag{3}$$

which implies $X = 0$ and thus satisfies the boundary conditions, Eq. (91–29). Hence this is an eigensolution, and $L = 0$ is an eigenvalue for all values of ω.

Connection with spheroidal wave functions. If $L \neq 0$, the elimination of Z results in the equation

$$\sigma^2 \frac{d^2 Y}{d\tau^2} + L^2(1 - a^2\tau^2) Y = 0 \tag{4}$$

for Y. The further substitution

$$Y = \sigma S \tag{5}$$

leads to

$$\frac{d}{d\tau}\left(\sigma^2 \frac{dS}{d\tau}\right) + [L^2(1 - a^2\tau^2) - 1/\sigma^2]S = 0. \tag{6}$$

This equation is a special case of the spheroidal wave equation,

$$\frac{d}{d\tau}\left(\sigma^2 \frac{dS_{ml}}{d\tau}\right) + [A_{ml} - h^2\tau^2 - m^2/\sigma^2]S_{ml} = 0 \tag{7}$$

which has been intensively studied.[(1),(4)] This equation reduces to Legendre's equation (Eq. (91–7)) if $h = 0$. The functions S_{ml} are therefore generalizations of P_l^m, and the notation of Ref. (4) has been devised to bring this into evidence. Thus

$$S_{ml} = S_{ml}(h,\tau) \tag{8}$$

are the eigensolutions of Eq. (95–7), and

$$S_{ml}(0,\tau) = P_l^m(\tau), \; A_{ml}(0) = l(l + 1). \tag{9}$$

Here l is always a positive integer, and

$$l = m, \quad m + 1, \quad m + 2 \ldots .$$

No matter what the value of h, S_{ml} has exactly $l - m$ roots for $|\tau| < 1$, and roots of degree $m/2$ at $\tau = \pm 1$. These properties are identical with those of P_l^m, and will be utilized below.

Comparing Eqs. (95–6) and (95–7), it is seen that

$$\begin{aligned} a^2L^2 &= h^2, \\ m &= 1, \\ L^2 &= A_{ml}(h). \end{aligned} \tag{10}$$

Tables for the calculation of $A_{ml}(h)$ are available,[4] so that it becomes possible to calculate associated values of $L = \varkappa r_S$ and $a = \Omega/\omega$ with h as parameter. A short table of such associated pairs follows, and is presented graphically on Fig. 79.

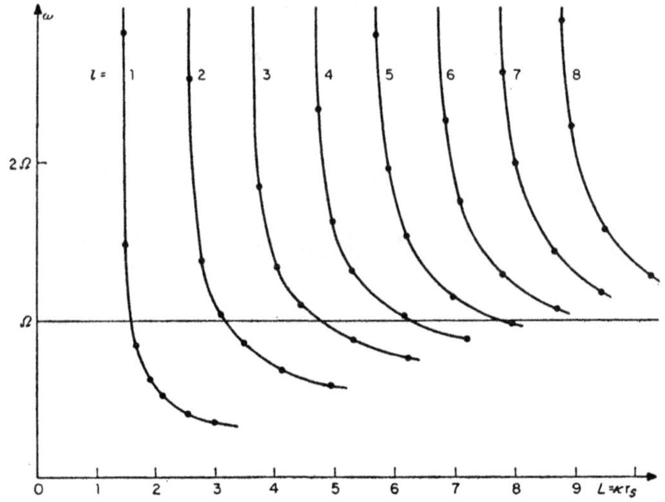

FIG. 79. Diagnostic diagram for the zonal oscillations.

TABLE 6

Associated values of h, $\varkappa r_S$ and ω/Ω

$l = 1$

h	$\varkappa r_S$	ω/Ω
0·0	1·414	∞
0·5	1·432	2·863
1·0	1·482	1·482
2·0	1·654	0·8267
3·0	1·872	0·6240
4·0	2·098	0·5244
5·0	2·313	0·4626
6·0	2·515	0·4192
7·0	2·704	0·3862
8·0	2·881	0·3601

$l = 2$

h	$\varkappa r_S$	ω/Ω
0·0	2·449	∞
0·5	2·471	4·942
1·0	2·535	2·535
2·0	2·766	1·383
3·0	3·091	1·030
4·0	3·457	0·8642
5·0	3·826	0·7653
6·0	4·183	0·6971
7·0	4·519	0·6456
8·0	4·835	0·6044

TABLE 6—*continued*

$l = 3$

h	$\varkappa r_S$	ω/Ω
0·0	3·464	∞
0·5	3·481	6·962
1·0	3·531	3·531
2·0	3·726	1·863
3·0	4·030	1·343
4·0	4·412	1·103
5·0	4·837	0·9674
6·0	5·275	0·8791
7·0	5·704	0·8149
8·0	6·116	0·7645

$l = 4$

h	$\varkappa r_S$	ω/Ω
0·0	4·472	∞
0·5	4·486	8·971
1·0	4·526	4·526
2·0	4·684	2·342
3·0	4·940	1·647
4·0	5·283	1·321
5·0	5·694	1·139
6·0	6·150	1·052
7·0	6·625	0·9465
8·0	7·101	0·8876

$l = 5$

h	$\varkappa r_S$	ω/Ω
0·0	5·477	∞
0·5	5·488	10·977
1·0	5·522	5·522
2·0	5·654	2·827
3·0	5·870	1·957
4·0	6·165	1·541
5·0	6·531	1·306
6·0	6·958	1·160
7·0	7·430	1·061
8·0	7·928	0·9910

$l = 6$

h	$\varkappa r_S$	ω/Ω
0·0	6·481	∞
0·5	6·490	12·980
1·0	6·519	6·519
2·0	6·631	3·316
3·0	6·817	2·272
4·0	7·073	1·768
5·0	7·394	1·479
6·0	7·776	1·296
7·0	8·213	1·173
8·0	8·693	1·087

$l = 7$

h	$\varkappa r_S$	ω/Ω
0·0	7·483	∞
0·5	7·492	14·983
1·0	7·516	7·516
2·0	7·615	3·807
3·0	7·777	2·592
4·0	8·001	2·000
5·0	8·285	1·657
6·0	8·626	1·438
7·0	9·019	1·288
8·0	9·461	1·183

$l = 8$

h	$\varkappa r_S$	ω/Ω
0·0	8·45	∞
0·5	8·493	16·985
1·0	8·514	8·514
2·0	8·601	4·301
3·0	8·746	2·915
4·0	8·945	2·236
5·0	9·199	1·840
6·0	9·504	1·584
7·0	9·859	1·408
8·0	10·261	1·283

Since $Y = \sigma S_{1l}$, it follows that Y has $l - 1$ roots for finite values of μ, and roots of first degree at $\tau = \pm 1$. According to the theory developed in Section 72, which applies to Eq. (95–4), these roots must be separated by l extrema. Since Eq. (95–2) is

$$\frac{dY}{d\tau} = -L^2 Z,$$

it follows that Z has exactly l non-polar roots. Consequently, the integer l, as defined by Eq. (95–10), is also the azimuthal ordinal number of the eigensolution. It can also be shown that the boundary conditions, Eq. (91–29), are satisfied.

The functions S_{ml} (h, τ). Just as the tables referred to[(4)] enable the calculation of the eigenvalues L, so they enable the computation of the infinite series that represent the functions S_{ml} themselves. It has been remarked above that the coefficients of these series are not easily calculated, so that this is no trivial matter. The task of plotting these functions, after the coefficients of the series are known, is still formidable and will not be attempted here. Instead, the methods of Chapter XIV will be used to obtain qualitative information concerning the relation between Y and Z.

The phase paths. Eqs. (95–1) and (95–2) are already in the canonic form, so that the theorems of Chapter XIV can be applied at once to the (Y, Z) graph. Typical cases are shown on Fig. 80. When $a^2 < 1$ (or $\omega^2 > \Omega^2$) the coefficients

$$\begin{aligned} A &= (1 - a^2\tau^2), \\ B &= L^2(1 - \tau^2), \end{aligned} \tag{21}$$

have no finite roots for any real value of μ; the signature is thus $(++)$ When $a^2 > 1$ (or $\omega^2 < \Omega^2$), the coefficient A has roots when

$$\tau = \tanh \mu = \pm 1/a = \pm \omega/\Omega.$$

The signature is therefore $(-+)$ $(++)$ $(-+)$. These results, combined with those concerning the roots of Y and Z obtained above, enable one to sketch the phase paths, as shown on Fig. 80.

It will be noted that the angular velocity of all these phase paths is positive, and that the total angular motion, from $\mu = -\infty$ to $\mu = +\infty$, is $l\pi$.

96. The solutions near the poles

When $m \neq 0$, the solutions of the tidal equations cannot be expressed in terms of known functions, except in very special cases. These special cases will be treated below, but a full understanding of the results requires some general information concerning the tidal equations and their solutions. In this section, the behaviour of the eigensolutions in very high latitudes will be investigated.

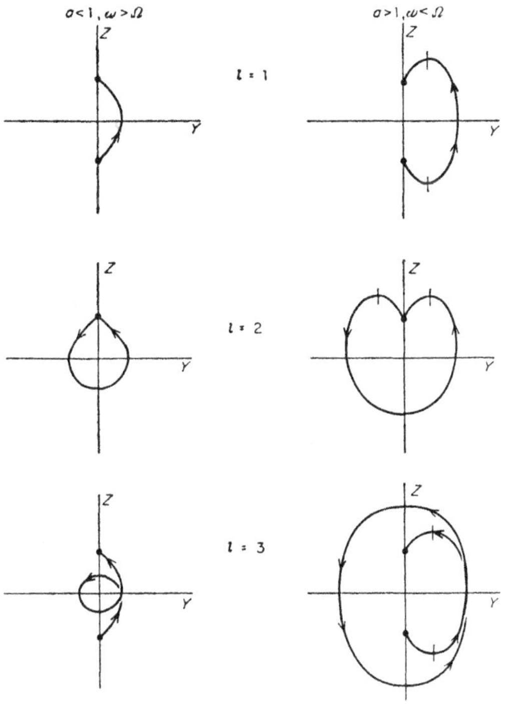

FIG. 80. Phase paths of the zonal oscillations.

It will be shown that both Y and Z vanish at the poles with a factor σ^m. This insures that the boundary conditions, Eq. (91–29), are satisfied if $m > 1$; the case $m = 1$ requires more detailed investigation.

Near the north pole, the variable

$$x = 1 - \tau \tag{1}$$

will have very small values; if x is used as independent variable, the tidal equations, Eq. (94–12, 13), become

$$
\begin{aligned}
-x(2-x)\frac{dZ}{dx} + sma(1-x)Z &= [1 - a^2(1-x)^2]Y, \\
-x(2-x)\frac{dY}{dx} - sma(1-x)Y &= [m^2 - L^2x(2-x)]Z.
\end{aligned}
\tag{2}
$$

The functions Y and Z may be expanded in series of powers of x:

$$
\begin{aligned}
Y &= x^\nu(a_0 + a_1x + \ldots), \\
Z &= x^\nu(b_0 + b_1x + \ldots),
\end{aligned}
\tag{3}
$$

where a_0, b_0 are not both zero. If these expansions are substituted into Eq. (96–2), and terms of the same degree in x are collected, the equations

$$
\begin{aligned}
&x^\nu[(sma - 2\nu)b_0 - (1 - a^2)a_0] + x^{\nu+1}[(\nu - sma)b_0 - 2a^2a_0 \\
&\qquad + (sma - 2\nu - 2)b_1 - (1 - a^2)a_1] + \ldots = 0, \\
&x^\nu[-(sma + 2\nu)a_0 - m^2b_0] + x^{\nu+1}[(\nu + sma)a_0 + 2L^2b_0 \\
&\qquad - (sma + 2\nu + 2)a_1 - m^2b_1] + \ldots = 0,
\end{aligned}
\tag{4}
$$

are obtained. The coefficients of each power of x must vanish separately; applying this principle to the νth power,

$$
\begin{aligned}
(1 - a^2)a_0 - (sma - 2\nu)b_0 &= 0, \\
(sma + 2\nu)a_0 + m^2b_0 &= 0.
\end{aligned}
\tag{5}
$$

Since a_0 and b_0 are not both zero, the determinant of these equations must vanish:

$$m^2(1 - a^2) + (m^2a^2 - 4\nu^2) = 0$$

whence

$$\nu = \pm m/2.$$

The negative sign is excluded for the eigensolutions since Y and Z do not become infinite; hence $\nu = m/2$ and one may take

$$a_0 = m, \quad b_0 = -(1 + sa). \tag{6}$$

Therefore, for small values of $(1 - \tau)$,

$$\begin{aligned} Y &= m(1 - \tau)^{m/2}[1 + \ldots], \\ Z &= -(1 + sa)(1 - \tau)^{m/2}[1 + \ldots]. \end{aligned} \tag{7}$$

The south polar regions can be treated in a similar manner, but symmetry considerations suffice to show that the eigensolutions are given by

$$\begin{aligned} Y &= \pm m(1 + \tau)^{m/2}[1 + \ldots] \\ Z &= \mp(1 + sa)(1 + \tau)^{m/2}[1 + \ldots] \end{aligned}$$

near the south pole. This may be summarized in the equations

$$\begin{aligned} Y &= m\sigma^m f(\tau), \\ Z &= -(1 + sa)\sigma^m g(\tau), \end{aligned} \tag{8}$$

where

$$\begin{aligned} f(1) &= g(1) = (1/2)^{m/2}, \\ f(-1) &= -g(-1) = \pm(1/2)^{m/2}. \end{aligned} \tag{9}$$

Eq. (94–9) then yields

$$X = \sigma^m[ma\tau f(\tau) - (1 + sa)smg(\tau)]. \tag{10}$$

It follows that, when $m = 1$, and $\tau = \pm 1$

$$\begin{aligned} X/\sigma &= \mp sg(\pm 1), \\ Y/\sigma &= f(\pm 1), \end{aligned}$$

and that the boundary conditions are all fulfilled even in this case; as well as when $m > 1$.

97. The tidal equations in canonic form

It has been noted that the tidal equations for $m = 0$ have the canonic form of Chapter XIV. When $m \neq 0$, the substitution

$$Y = \sigma^{-sma}y, \quad Z = m^2\sigma^{sma}z, \tag{1}$$

converts Eqs. (94–12, 13) into the canonic pair

$$\frac{dz}{d\mu} = (m^2 - b^2\tau^2)\sigma^{-2sb}y, \tag{2}$$

$$\frac{dy}{d\mu} = -\left(\frac{L^2}{m^2}\sigma^2 - 1\right)\sigma^{+2sb}z, \tag{3}$$

where

$$b = ma. \tag{4}$$

In all the following applications, m will have only the values 1, 2, 3 . . . . However, here it will be necessary to consider more general values of m, and, in particular, sometimes to consider b as a constant, while m and L are allowed to vary continuously.

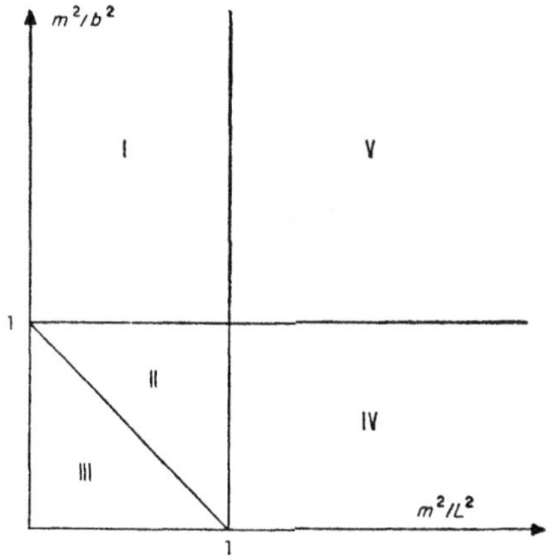

FIG. 81. The five special cases of the tidal equation.

The signatures. For any fixed value of b, one may construct a diagram, analogous to the $(\varkappa, \omega)$ diagram of Chapter XIV, with m^2/L^2 and m^2/b^2 as rectangular coordinates: see Fig. 81. The first quadrant of this diagram is divided into five areas, and solutions corresponding to any point of one of these areas will have the same signature. These signatures are

Area I:		$(+-)$	$(++)$	$(+-)$	
Area II:	$(--)$	$(+-)$	$(++)$	$(+-)$	$(--)$
Area III:	$(--)$	$(-+)$	$(++)$	$(-+)$	$(--)$
Area IV:		$(--)$	$(+-)$	$(--)$	
Area V:			$(+-)$		

The phase paths. The variables y and z may be plotted with μ (or τ) as parameter, yielding phase paths of the sort described in Chapter XIV. It will be noted that the signatures listed above are independent of the quantity s and of the magnitude of b (or a); this is not true of the phase paths. Eq. (97–1) involves both s and

a, and these quantities will determine the character of the phase path for large values of μ. Referring to Eq. (96–8), it is seen that

$$\begin{aligned} y &= m\sigma^{m(1+sa)}f(\tau), \\ z &= (1/m^2)\sigma^{m(1-sa)}g(\tau). \end{aligned} \tag{5}$$

As $\mu \to \infty$, $\sigma \to 0$ and y will become zero if $1 + sa > 0$, or infinite if $1 + sa < 0$. Similarly z will become zero if $1 - sa > 0$, and infinite

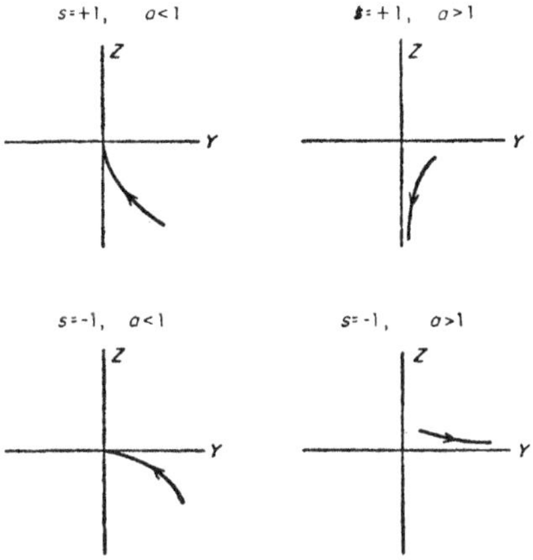

FIG. 82. Asymptotic behaviour of the eigensolutions.

if $1 - sa < 0$. It follows that those parts of the phase paths that correspond to large positive values of μ may have any one of the four characters illustrated by Fig. 82. The character of the paths for large negative values of μ can be deduced from these by considerations of symmetry.

Using this information about the beginning and end of the phase paths, it is readily shown that there can be no eigensolutions in area V, whose signature is $(+-)$. The qualitative character of the phase paths corresponding to the other four areas can also be sketched in conformity with the theorems of Chapter XIV. Such sketches are shown on Figs. 83 and 84.

It is seen that the total angle swept out by the radius vector is always an integral multiple of π; call this integer p. In areas II and III, p may be positive, negative, or zero. In area I, $p = 0, 1, 2 \ldots$ if $s = +1$, but when $s = -1$, $p = 0$ is impossible, and $p = 1, 2, 3 \ldots$. In area IV, $p = 0, -1, -2, \ldots$ if $s = -1$, but when $s = +1$, $p = 0$ is impossible, and $p = -1, -2, -3 \ldots$.

The number of roots of z (other than those for $\mu = \pm\infty$) may also be counted: this has already been called k in Chapter XVI. However, it is now necessary to take account of the direction in which the phase path crosses the y-axis, and to redefine k as the algebraic sum of these crossings (cf. Sections 87, 88). Then Figs. 83 and 84 show that $k = p$

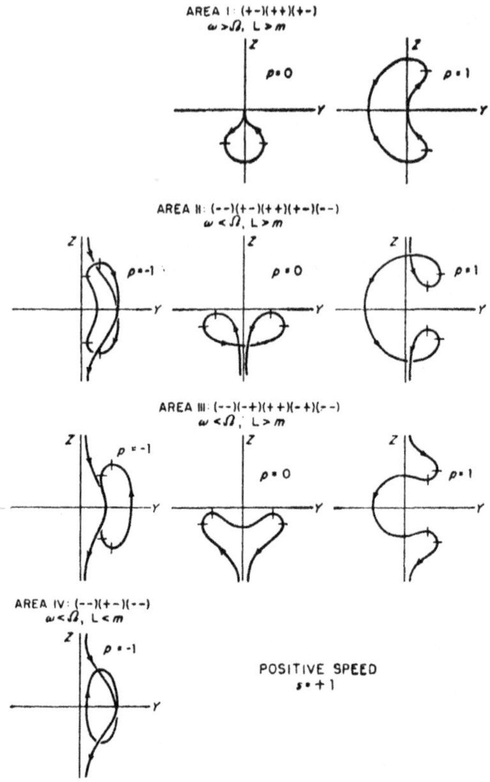

FIG. 83. Phase paths of eigensolutions for positive speed, $s = +1$.

when $s = +1$. However, when $s = -1$, the situation is more complicated; in area I, $k = p - 1$, while in areas II, III and IV, $k = p + 1$.

The modal curves. Those points in the (m^2/L^2), (m^2/b^2) diagram that represent eigensolutions will again lie on curves—the modal curves—as was found to be the case in Chapter XIII. Let $\vartheta(\mu, m, L, b)$ be the polar angle of the phase path, defined so that $\vartheta(-\infty, m, L, b) = 0$. Then the equation of the modal curve will be

$$\vartheta(+\infty, m, L, b) = p\pi, \tag{6}$$

p being the integer defined above.

Sturm's formula (Section 84) will now be used to prove certain properties of the modal curves. Let (y_1, z_1) and (y_2, z_2) be two eigensolutions that correspond to the same values of b and p, but different

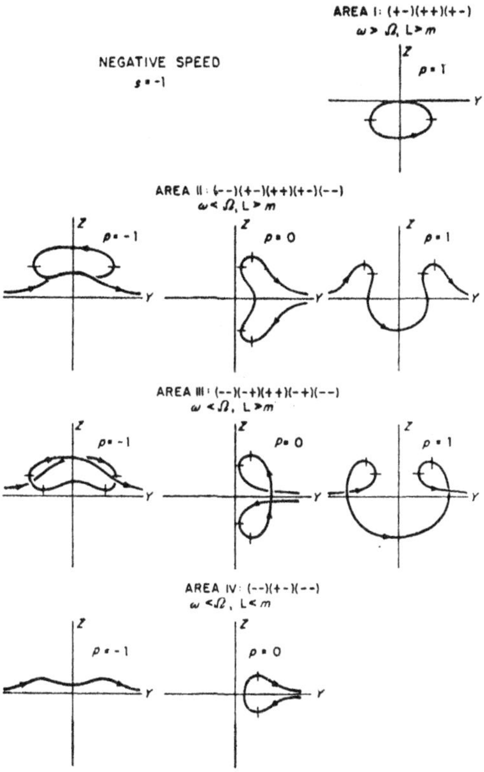

FIG. 84. Phase paths of eigensolutions for negative speed, $s = -1$.

values of m and L. Then, in the notation of Eqs. (97–2) and (97–3), Sturm's formula is

$$\frac{\mathrm{d}}{\mathrm{d}\mu}(z_1 y_2 - y_1 z_2) = (m_1^{\,2} - m_2^{\,2})\sigma^{-2sb} y_1 y_2 + \left(\frac{L_1^{\,2}}{m_1^{\,2}} - \frac{L_2^{\,2}}{m_2^{\,2}}\right)\sigma^{2sb+2} z_1 z_2. \quad (7)$$

As in Section 84, one concludes from this equation that $\vartheta(\mu, m, L, b)$ is a monotonically increasing function of both m^2 and L^2/m^2 when μ and b are held constant. In turn, it follows that the graph of Eq.

(97–6), in the (m^2/L^2), (m^2/b^2) plane will everywhere have a positive slope; moreover, that as the integer increases, the modal curves move upward and to the left.

Combining everything that has been learned thus far concerning the modal curves and the integer p, it is reasonable to suppose that the completed diagnostic diagram will have somewhat the appearance

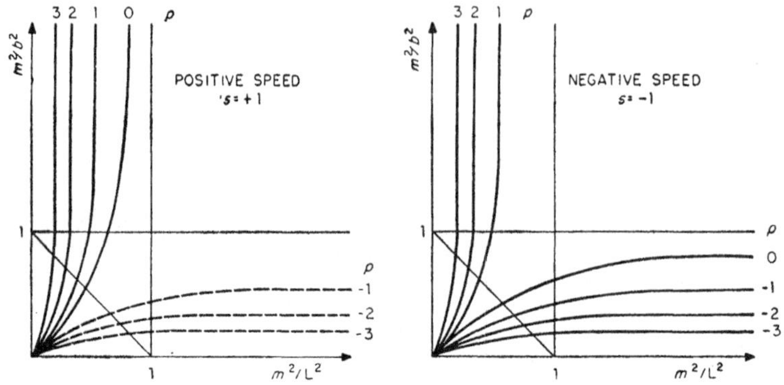

FIG. 85. Diagnostic diagrams in the (m^2/b^2), (m^2/L^2) plane.

of Fig. 85. It should be noted that the cases $s = \pm 1$ differ in the location of the mode $p = 0$. However reasonable it may appear, this conjecture is only a working hypothesis, to be tested and corrected below.

Transformation to the $(\varkappa, \omega)$ plane. In constructing the Fig. 85, it has been supposed that $b = ma = m\Omega/\omega$ is constant, while m and $L = \varkappa r_S$ are continuously variable. The plane diagrams of Fig. 85 are thus sections of a three-dimensional space, b being the third co-ordinate. In testing the validity of the above conjecture, diagrams like Fig. 86 will be more convenient in which $m =$ constant and ω and $L = \varkappa r_S$ are the variable co-ordinates. These $(\varkappa, \omega)$ diagrams are essentially other sections of the same three-dimensional figure just mentioned, with the further difference that L, rather than m^2/L is used as one cartesian co-ordinate. It is not difficult to see, qualitatively, how these other sections will appear. Fig. 86 presents the same information as Fig. 85, but in this new co-ordinate system, m being held constant, while ω and $\varkappa r_S$ vary continuously. Fig. 86 should be compared with Fig. 79, the latter being quantitatively accurate for $m = 0$. In this special case $(m = 0)$, the quantity s disappears from the equations. It also follows from the theory of the spheroidal wave equation that there are no eigensolutions with negative p; in fact, this conclusion follows directly from the fact that the signature of all solutions is then $(-+)$ $(++)$ $(-+)$ (cf. the discussion of the phase paths in Section 95).

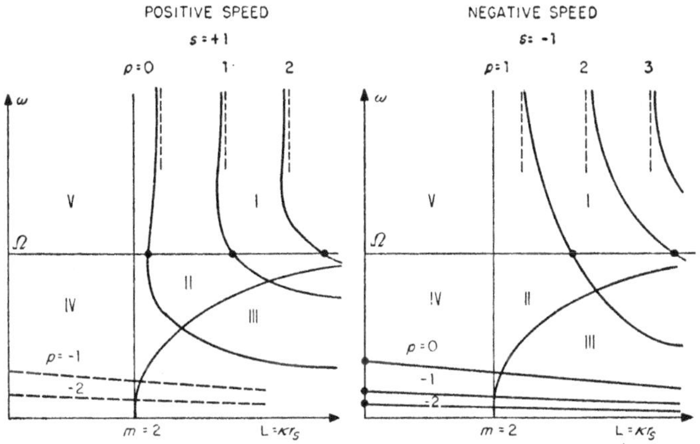

FIG. 86. Diagnostic diagrams in the $\varkappa$, ω phase.

98. The high-frequency limit

The last two sections were devoted to a general survey of the tidal equations and their eigensolutions. The results obtained were often qualitative; it remains to deduce those quantitative results that are obtainable without great labour. The first set of these relates to the limit $a \to 0$, or $\omega \to \infty$. It has already been noted, in Section 94, that $a = 0$ also is equivalent to $\Omega = 0$; in the high-frequency limit, the equations and their eigensolutions will approach those discussed in Chapter XVI. The nature of this limiting approach requires some further study, however.

When $a = 0$, the eigensolutions are

$$\begin{aligned} Z &= P_l^m(\tau), \\ Y &= \sigma^2 \, \mathrm{d}P_l^m/\mathrm{d}\tau, \\ X &= smP_l^m(\tau). \end{aligned} \tag{1}$$

and the eigenvalues are

$$L^2 = l(l+1), \qquad l = m, m+1, \ldots . \tag{2}$$

The phase diagrams of some of these eigensolutions are shown on Fig. 77 of Chapter XVI.

The problem is, to enter these quantitative results on the $(\varkappa, \omega)$ diagram, Fig. 86. In order to do so, it is necessary to determine the value of p that is associated to a given set of values of l and m. This is quite simple in a formal way. It has been noted in Chapter XVI, that the number of roots of Z is $k = l - m$. When a is large, and $L^2 > m^2$, the point $(\varkappa, \omega)$ lies in area I of the diagnostic diagram. Finally, in Section 97, it was noted that, in this area, $k = p$ when $s = +1$, and $k = p - 1$ when $s = -1$: hence

$$\begin{aligned} p &= l - m & \text{when } s &= 1, \\ &= l - m + 1 & \text{when } s &= -1. \end{aligned} \tag{3}$$

In Fig. 86, m is fixed at the value 2, by way of example. Hence for $s = +1$, the curves for $p = 0, 1, 2$ should approach the asymptotes

$$\begin{aligned} L^2 &= l(l + 1) = (p + m)(p + m + 1) \\ &= (p + 2)(p + 3) \end{aligned}$$

as $\omega \to \infty$. These asymptotic values have been indicated by the dotted vertical lines. Similarly, for $s = -1$, the curves for $p = 1$, $2 \ldots$ should have asymptotes at

$$\begin{aligned} L^2 &= l(l + 1) = (p + m - 1)(p + m) \\ &= (p + 1)(p + 2), \end{aligned}$$

and these have also been indicated on Fig. 86.

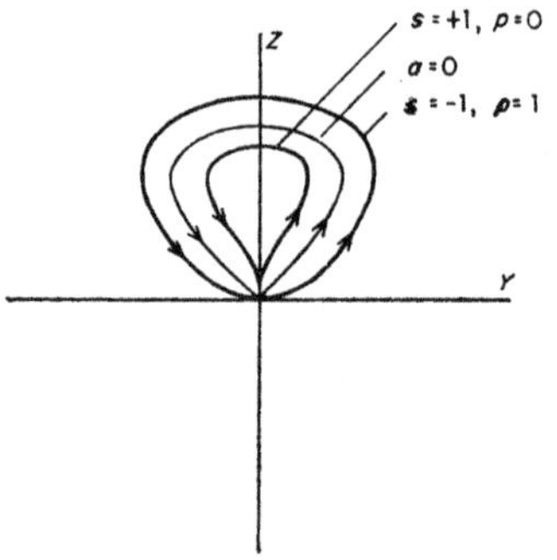

FIG. 87. Phase paths at the high frequency-limit.

However, a paradox becomes apparent on inspection of Fig. 77. The integer p is defined in terms of the total angle swept out by the radius vector to the phase point, as the latter describes its whole path.

According to Section 97, this angle should be $p\pi$; but the phase paths of Fig. 77 show that this angle is not an integral multiple of π (except for $m = 0$).

This paradox is resolved by noting that Fig. 77 represents the phase paths for $a = 0$, while the present discussion concerns small, but still positive, values of a. In Fig. 87, the light curve reproduces the case $l = 2$, $m = 2$, $k = 0$ illustrated on Fig. 87. The two heavy curves sketch the phase path for a small, non-zero value of a, and for $s = \pm 1$. The heavy curves correspond to $p = 0$ and 1 respectively. As $a \to 0$, they will both approach the light curve, for which p is not an integer, but the convergence will not be uniform in μ.

99. The semi-diurnal oscillations

A second opportunity for quantitative results is afforded by the case $\omega = \Omega$, or $a = 1$. The tidal equations (Eqs. 94–12, 13) then become

$$\begin{aligned} \frac{dZ}{d\mu} + sm\tau Z &= \sigma^2 Y, \\ \frac{dY}{d\mu} - sm\tau Y &= -(L^2\sigma^2 - m^2)Z. \end{aligned} \tag{1}$$

If $m = 0$, these simplify still further, and can be solved in terms of elementary functions. If $m \neq 0$, the elimination of Y yields the second-order equation

$$\sigma^2 \frac{d^2 Z}{d\tau^2} + [L^2\sigma^2 - sm - (m^2 - 2sm)/\sigma^2]Z = 0. \tag{2}$$

The substitution

$$Z = \sigma S \tag{3}$$

reduces this to the spheroidal wave equation

$$\frac{d}{d\tau}\left(\sigma^2 \frac{dS}{d\tau}\right) + [(L^2 - m^2) - L^2\tau^2 - (m - s)^2/\sigma^2]S = 0. \tag{4}$$

The identification with the notation of Ref. (4) is now

$$S = S_{m'l'}(h, \tau), \tag{5}$$

$$m' = m - s \tag{6}$$

$$h^2 = L^2 \tag{7}$$

$$L^2 - sm = A_{m'l'}(h). \tag{8}$$

These equations combine to yield

$$L^2 - A_{m'l'}(L) = sm' + 1, \tag{9}$$

which is a transcendental equation determining the eigenvalue L. Using the tables previously mentioned,(4) it can be solved graphically with considerable precision. The results of these calculations are

TABLE 7

Values of $\varkappa r_S$ for semi-diurnal oscillations

Positive speed, $s = +1$

l'	m'	$L = \varkappa r_S$	m	$k = l' - m' = p$
0	0	1·297	1	0
1	0	2·589	1	1
	1	2·209	2	0
2	0	4·088	1	2
	1	3·597	2	1
	2	3·209	3	0
3	0	5·622	1	3
	1	5·068	2	2
	2	4·602	3	1
	3	4·209	4	0
4	0	7·171	1	4
	1	6·577	2	3
	2	6·509	3	2
	3	5·605	4	1
	4	5·209	5	0

Negative speed, $s = -1$

l'	m'	$L = \varkappa r_S$	m	$k = l' - m' = p - 1$
2	2	2·401	1	0
3	2	3·968	1	1
	3	3·336	2	0
4	2	5·534	1	2
	3	4·879	2	1
	4	4·303	3	0

given in the accompanying table, which lists not only the values of $L = \varkappa r_S$ for various values of l' and m', but also the values of m, k and p.

The integer $k = l' - m'$ is the number of non-polar roots of Z; this follows from known properties of the spheroidal wave functions. The determination of p presents no difficulties when $s = +1$, since then $k = p$ in all four areas of the diagnostic diagram (cf. Section 97). However, these points lie on the boundary between areas I and II, and when $s = -1$, $p = k + 1$ in area I and $p = k - 1$ in area II. The proper assignment is thus problematic.

The problem finds a simple solution when the phase paths are sketched. Fig. 88 shows three such paths, for $s = -1$, $p = 1$, and $a < 1$, $= 1$, and >1, respectively. It is at once clear that $p = k + 1$ for $a \leqslant 1$, and $p = k - 1$ for $a > 1$.

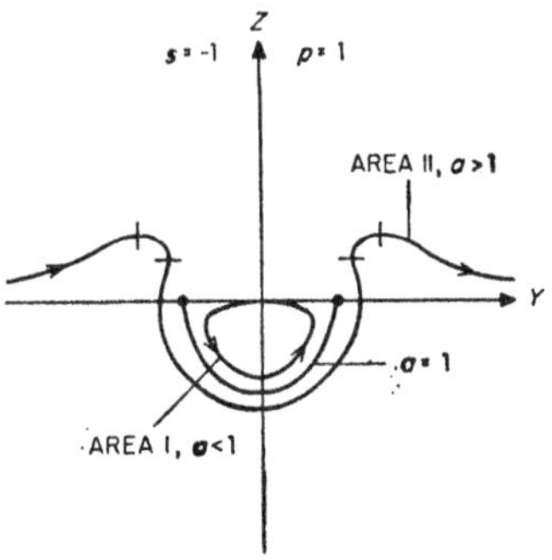

FIG. 88. Phase paths for oscillations whose period is nearly 12 hours.

Having identified the proper values of p, the data of the foregoing table provide points for Fig. 86; they are shown by the circles at $\omega = \Omega$.

100. Oscillations of the second kind, and the "long waves"

One other special case of the tidal equations can be solved in terms of known functions: this is obtained by setting $L = 0$. It is of particular interest because it is the only solvable example of a class of eigensolutions that Laplace called "oscillations of the second kind." He characterized them by the fact that their frequency, ω, approaches zero as Ω approaches zero. In the present notation, they are the eigensolutions for $s = 1, p < 0$ or $s = -1, p \leqslant 0$ (cf. Fig. 86). They present a number of problems that have not yet been solved.

When $L = 0$, the tidal equations become

$$\frac{dZ}{d\mu} + sma\tau Z = (1 - a^2\tau^2)Y, \tag{1}$$

$$\frac{dY}{d\mu} - sma\tau Y = m^2 Z. \tag{2}$$

This has become known as the equation for "long" waves. After eliminating Z, it is seen that Y satisfies Legendre's equation.

$$\frac{d}{d\tau}\left(\sigma^2 \frac{dY}{d\tau}\right) + [-sma - m^2/\sigma^2]Y = 0. \tag{3}$$

The case $m = 0$ has already been included above, so that it may be supposed that $m > 0$.

The eigensolutions. As discussed in Chapter XVI, this equation has eigensolutions only when

$$-sma = l'(l' + 1), \qquad l' = m, m + 1, \ldots. \tag{4}$$

Since m, a and l' are all positive, it follows that Eq. (100–3) has no eigensolutions for $s = +1$, but only for $s = -1$. Equation (100–4) may then be written, since $a = \Omega/\omega$,

$$\omega = m\Omega/l(l + 1). \tag{5}$$

It is clear that this conforms to Laplace's definition of the oscillations of the second kind.

The eigensolutions corresponding to Eq. (100–4) are

$$\begin{aligned} Y &= P_{l'}^{m}(\tau), \\ m^2 Z &= (l' + m)P_{l'}{}^{m}{}_{-1} + l'^2 \tau P_{l}{}^{m}. \end{aligned} \tag{6}$$

This expression for Z follows from Eq. (100–2) and Eq. (91–12). The first few of these eigensolutions are listed below.

TABLE 8

Eigensolutions

l'	m	Y	Z
1	1	$-\sigma$	$-\sigma\tau$
2	2	$3\sigma^2$	$3\sigma^2\tau$
	1	$-3\sigma\tau$	$-3\sigma(1 + 4\tau^2)$
3	3	$-15\sigma^3$	$-15\sigma^3\tau$
	2	$15\sigma^2\tau$	$15\sigma^2(1 + 9\tau^2)/4$
	1	$-3(5\tau^2 - 1)\sigma/2$	$-3\sigma(45\tau^3 - \tau)/2$

The phase paths. In order to discuss the phase paths for these eigensolutions, it is necessary to calculate the canonic variables, y and z, according to Eq. (97–1). Their functional forms are listed in the following table. It is seen that these functions correspond to the sketches of Fig. 84, with the values of k and p indicated in the table. Since these values of p are all zero or negative, the eigenvalues given by Eq. (100–5) confirm the sketch of Fig. 86 for $s = -1$; these values are indicated by the circles on the ω-axis of that sketch.

Positive speed. The existence of the oscillations of the second kind having thus been confirmed by example in the case $s = -1$, the question of their existence for $s = +1$ is raised. Their existence is not excluded by the general considerations of Section 97, but, to the best of the writer's knowledge, no one has succeeded in exhibiting any examples of these solutions. While this may make their existence improbable, it does not constitute a proof of non-existence.

TABLE 9

The functions y and z

l'	m	y	z	k	p
1	1	$-1/\sigma$	$-\sigma^3\tau$	1	0
2	2	$3/\sigma^4$	$3\sigma^8\tau/4$	1	0
	1	$-3\tau/\sigma^5$	$-3\sigma^7(1 + 4\tau^2)$	0	-1
3	3	$-15/\sigma^9$	$-15\sigma^{15}\tau/9$	1	0
	2	$15\tau/\sigma^{10}$	$15\sigma^{14}(1 + 9\tau^2)/16$	0	-1
	1	$-3(5\tau^2 - 1)/2\sigma^{11}$	$-3\sigma^{13}(45\tau^3 - \tau)/2$	-1	-2

19

Perhaps the following can be accepted as such a proof: if they exist, their modal curves must have a positive slope in the m^2/L^2, m^2/b^2 diagram, and must lie entirely in the areas below the line $\omega = \Omega$. In the $(\varkappa, \omega)$ diagram, they must therefore intersect the ω-axis for finite values of ω. Since the ω-axis is the line $L = 0$, the eigensolutions for these values of ω must satisfy Eq. (100–3). It has been seen that this equation has no eigensolutions for $s = +1$. This is a contradiction. Hence, either there are no eigensolutions for $s = +1$, or else there is some inobvious presupposition hidden in the foregoing analysis of the problem.

For example: it has been tacitly assumed that $\varkappa$ is real—i.e., that $L^2 \geqslant 0$. In Chapter X, it was seen that this is not necessarily the case. When $L^2 < 0$, the signature of the solutions is $(--)\ (+-)\ (--)$, and they have the same properties as in area IV. However, it appears that even in this case, the tidal equations have eigensolutions only for $s = -1$; at least, no examples for $s = +1$ have been found by the writer.

The Rossby waves. C.-G. Rossby[3] considers the periodic disturbances of a uniform zonal stream on a rotating planet. Let u_0 be the velocity of this stream, positive when it flows from west to east. The dimensionless constants

$$\beta = \Omega\sigma/r_S, \qquad g = \Omega\tau/(gD_0)^{\frac{1}{2}} \tag{7}$$

are useful, D_0 being the "mean depth of the uniform atmosphere". Rossby[3] then introduces various simplifying assumptions, and deduces that the phase velocity of the wave is

$$c_{\text{ph}} = \frac{u_0\varkappa^2 - \beta}{\varkappa^2 + q^2}. \tag{8}$$

When u_0 is zero, $c_{\text{ph}} < 0$—i.e. the waves are propagated only from east to west. This fact suggests that they are identical with the oscillations of the second kind.

B. Haurwitz[2] gives a somewhat more elaborate treatment of the Rossby waves. He supposes that the velocity of the zonal stream is

$$u_0 = \omega_0 r_S \cos\phi = \omega_0 r_S \sigma, \tag{9}$$

where ω_0 is a constant angular velocity and ϕ the latitude. After correcting several obvious misprints, Eq. (129) of Ref. (2) may be written

$$\frac{sm(\Omega + 2\omega_0)}{m\omega_0 - \omega} = l'(l' + 1). \tag{10}$$

When $\omega_0 = 0$, this equation becomes identical with Eq. (100–4) above. There is therefore little doubt that the Rossby waves are essentially identical with Laplace's oscillations of the second kind. Moreover, it is clear that they exist even in static atmospheres and then have nothing to do with zero-order motion. Neither is Haurwitz's assumption, that $L = 0$, an essential element in the definition of these waves.

References for Chapter XVII

1. Carson Flammer, *Spheroidal Wavefunctions*, Stanford Univ. Press, 1957.
2. B. Haurwitz, The perturbation equations in meteorology, p. 417 of *Compendium of Meteorology*, American Meteorological Society, Boston, 1951.
3. C.-G. Rossby, On the dispersion of planetary waves in a barotropic atmosphere, *Tellus*, **1**, 1 (1949).
4. Stratton, Morse, Chu, Little, and Corbató, *Spheroidal Wave Functions*, Technology Press of M.I.T., Cambridge, Mass., 1956.

APPENDIX:

MERCATOR CO-ORDINATES

On a sphere of unit radius, the element of length is dS, where

$$dS^2 = \cos^2 \phi \, d\lambda^2 + d\phi^2 \tag{1}$$

λ being the longitude and ϕ the latitude. The vertical co-ordinate, μ, on a Mercator chart of the sphere is defined by

$$d\phi = \cos \phi \, d\mu, \tag{2}$$

so that

$$dS^2 = \cos^2 \phi(d\lambda^2 + d\mu^2).$$

Equation (2) is readily integrated to yield the four equivalent equations

$$\left.\begin{aligned} \sin \phi &= \tanh \mu, \\ \cos \phi &= \operatorname{sech} \mu, \\ \tan \phi &= \sinh \mu, \\ \sec \phi &= \cosh \mu, \end{aligned}\right\} \tag{3}$$

so that finally,

$$dS^2 = \operatorname{sech}^2 \mu(d\lambda^2 + d\mu^2) \tag{4}$$

In terms of Gudermann's function, Eqs. (3) may be written

$$\phi = gd \; \mu. \tag{5}$$

A short table of the gudermannian follows: a more extensive table is to be found in the *Smithsonian Mathematical Tables—Hyperbolic Functions*, by Becker and Van Orstrand (Washington, 1909). Fig. 89 shows the relation between ϕ and μ graphically.

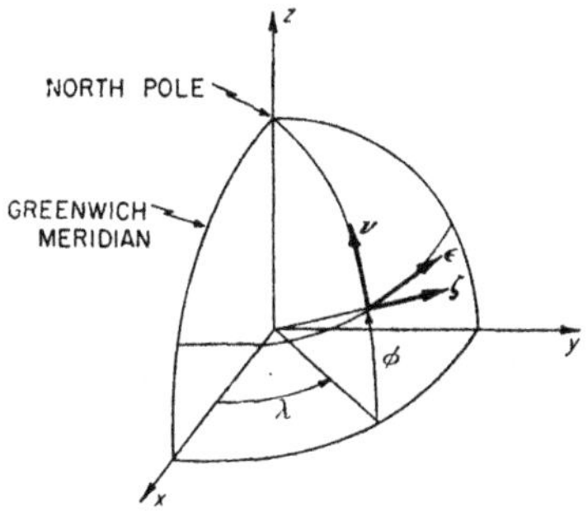

FIG. 89. The unit vectors **ε**, **ν**, **ζ** and their relation to latitude and longitude.

TABLE 10

Table of the gudermannian function

Degrees		Radians	
μ	ϕ	μ	ϕ
0	0	0·0	0·000
10·05	10	0·2	0·199
20·45	20	0·4	0·390
31·5	30	0·6	0·567
43·7	40	0·8	0·726
57·8	50	1·0	0·865
75·4	60	1·2	0·986
99·4	70	1·4	1·087
139·6	80	1·6	1·172
∞	90	1·8	1·243
		2·0	1·302
		2·2	1·350
		2·4	1·390
		2·6	1·423
		2·8	1·449
		3·0	1·471
		3·5	1·510
		4·0	1·534
		5·0	1·557
		6·0	1·566

When combined with the radius vector, r, the Mercator co-ordinates can be used in place of polar co-ordinates. The explicit form of the transformation to the cartesian system is

$$\begin{aligned} x &= r \operatorname{sech} \mu \cos \lambda, \\ y &= r \operatorname{sech} \mu \sin \lambda, \\ z &= r \tanh \mu. \end{aligned} \tag{6}$$

Unit vectors in the east, north, and zenith directions are

$$\begin{aligned} \boldsymbol{\epsilon} &= -\mathbf{i} \sin \lambda + \mathbf{j} \cos \lambda, \\ \boldsymbol{\nu} &= -\mathbf{i} \cos \lambda \tanh \mu - \mathbf{j} \sin \lambda \tanh \mu + \mathbf{k} \operatorname{sech} \mu, \\ \boldsymbol{\zeta} &= \mathbf{i} \cos \lambda \operatorname{sech} \mu + \mathbf{j} \sin \lambda \operatorname{sech} \mu + \mathbf{k} \tanh \mu. \end{aligned} \tag{7}$$

Since these vectors are orthogonal,

$$\begin{aligned} \boldsymbol{\epsilon}^2 = \boldsymbol{\nu}^2 = \boldsymbol{\zeta}^2 = 1, \quad \boldsymbol{\epsilon} \cdot \boldsymbol{\nu} = \boldsymbol{\nu} \cdot \boldsymbol{\zeta} = \boldsymbol{\zeta} \cdot \boldsymbol{\epsilon} = 0, \\ \boldsymbol{\epsilon} \times \boldsymbol{\nu} = \boldsymbol{\zeta}, \quad \boldsymbol{\nu} \times \boldsymbol{\zeta} = \boldsymbol{\epsilon}, \quad \boldsymbol{\zeta} \times \boldsymbol{\epsilon} = \boldsymbol{\nu}. \end{aligned} \tag{8}$$

Any vector can be resolved into components with respect to these unit vectors:

$$\mathbf{F} = \boldsymbol{\epsilon} F_\epsilon + \boldsymbol{\nu} F_\nu + \boldsymbol{\zeta} F_\zeta, \tag{9}$$

and the formulae of vector algebra are the same in terms of these components as in terms of cartesian.

However, since the vectors $\boldsymbol{\epsilon}$, $\boldsymbol{\nu}$, $\boldsymbol{\zeta}$ depend on the co-ordinates λ, μ, r, the formulae for the operator ∇ are quite different. The derivatives of these vectors are given in tabular form below.

TABLE 11

Derivatives of the unit vectors

	$\boldsymbol{\epsilon}$	$\boldsymbol{\nu}$	$\boldsymbol{\zeta}$
$\dfrac{\partial}{\partial \lambda}$	$\boldsymbol{\nu} \tanh \mu - \boldsymbol{\zeta} \operatorname{sech} \mu$	$-\boldsymbol{\epsilon} \tanh \mu$	$\boldsymbol{\epsilon} \operatorname{sech} \mu$
$\dfrac{\partial}{\partial \mu}$	0	$-\boldsymbol{\zeta} \operatorname{sech} \mu$	$\boldsymbol{\nu} \operatorname{sech} \mu$

From this and Eqs. (6), it follows that

$$\nabla\psi = m\left(\boldsymbol{\epsilon}\frac{\partial\psi}{\partial\lambda} + \boldsymbol{\nu}\frac{\partial\psi}{\partial\mu}\right) + \boldsymbol{\zeta}\frac{\partial\psi}{\partial r}, \tag{10}$$

$$\nabla^2\psi = m^2\left(\frac{\partial^2\psi}{\partial\lambda^2} + \frac{\partial^2\psi}{\partial\mu^2}\right) + \frac{1}{r^2}\frac{\partial}{\partial r}\left(r^2\frac{\partial\psi}{\partial r}\right), \tag{11}$$

where

$$m = (1/r)\cosh\mu. \tag{12}$$

Also

$$\nabla\cdot\mathbf{F} = m\left[\frac{\partial F_\epsilon}{\partial\lambda} + m\frac{\partial}{\partial\mu}\left(\frac{F_\nu}{m}\right)\right] + \frac{1}{r^2}\frac{\partial}{\partial r}(r^2F_\zeta), \tag{13}$$

and

$$\nabla\times\mathbf{F} = \boldsymbol{\epsilon}\left[m\frac{\partial F_\zeta}{\partial\mu} - \frac{1}{r}\frac{\partial}{\partial r}(rF_\nu)\right] + \boldsymbol{\nu}\left[\frac{1}{r}\frac{\partial}{\partial r}(rF_\epsilon) - m\frac{\partial F_\zeta}{\partial\lambda}\right] + \boldsymbol{\zeta}\left[m\frac{\partial F_\nu}{\partial\lambda} - m^2\frac{\partial}{\partial\mu}\left(\frac{F_\epsilon}{m}\right)\right], \tag{14}$$

so that

$$\mathbf{M} = \boldsymbol{\zeta}\times\nabla = m\left(-\boldsymbol{\epsilon}\frac{\partial}{\partial\mu} + \boldsymbol{\nu}\frac{\partial}{\partial\lambda}\right). \tag{15}$$

It is frequently necessary to obtain the equation of a great circle on the unit sphere. Since the circle is the intersection of the sphere with a plane through its centre, this problem is the equivalent of finding the equation of such a plane. The plane can be specified by two angles, λ_0 and ϑ. The line OP is the intersection of the plane OPQ with the equatorial plane, $z = 0$, and λ_0 is the angle between OP and the positive x-axis. The plane OPQ makes the dihedral angle ϑ with the plane $z = 0$. Then the unit normal to OPQ is

$$\mathbf{n} = \mathbf{i}\sin\vartheta\sin\lambda_0 - \mathbf{j}\sin\vartheta\cos\lambda_0 + \mathbf{k}\cos\vartheta$$

and the equation of the plane is

$$r\{-\sin\vartheta\,\mathrm{sech}\,\mu\sin(\lambda-\lambda_0) + \cos\vartheta\tanh\mu\} = 0.$$

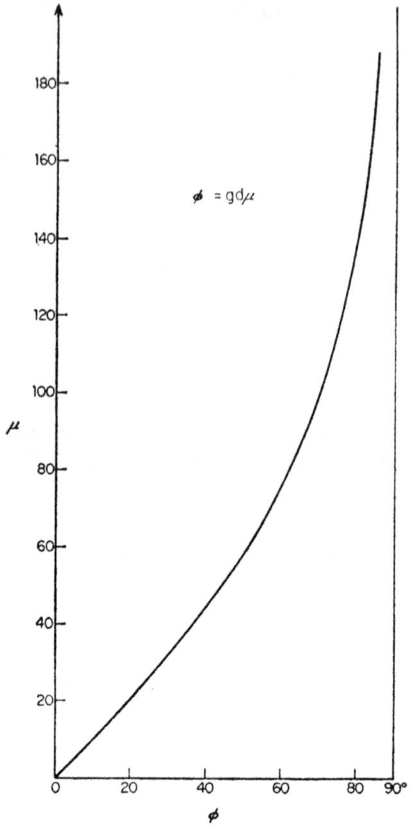

FIG. 90. The gudermannian function.

Dividing by r sech μ cos ϑ, this may be written

$$\sinh \mu = \tan \vartheta \sin (\lambda - \lambda_0).$$

Defining

$$\sinh \mu_0 = \tan \vartheta$$

this is

$$\sinh \mu = \sinh \mu_0 \sin (\lambda - \lambda_0). \tag{16}$$

INDEX

Printed in Dunstable, United Kingdom

91342994R00172